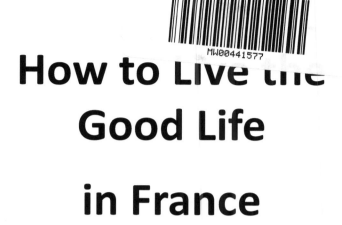

How to Live the Good Life

Good Life

in France

Attaining Your French Rural Dream

Lorraine Turnbull

Copyright© 2020

Cover design: bookscovered.co.uk

Cover photo © Lorraine Turnbull

Fat Sheep Press, Le Bois Vert, Lescarpedie, 24220 Meyrals, France

British Library Cataloguing in Publication Data

A CIP record for this book is available from the British Library

ISBN 978-9163890-1-4

For my family

Testimonials

"An absolute 'must buy' for anyone seriously thinking of moving to live in the French countryside. Brexit ready and full of information and best of all written by someone who has actually done it."   Brian McMillan

"The perfect kick up the backside I needed to go and do what I've dreamed of for years. Thanks for writing this as a follow up to The Sustainable Smallholders Handbook"   Susan Miller

"Lorraine Turnbull has written about the reality of rural life in France. Armed with this book I feel able to seriously think about moving to a country with a slower pace of life and better weather!"   Robert Richards

"I really liked the case studies from all over France written by people who have actually moved to live there and even set up their own rural businesses. Although the book is stuffed full of detail, Lorraine writes in an easy to understand, friendly style, so it feels like a friend is advising me."   Reader (Amazon)

For other testimonials please see the entry for the book on the Amazon website.

Lorraine Turnbull has also written The Sustainable Smallholders' Handbook (2019)

Connect with Lorraine:

Sustainable Smallholding Facebook page
https://www.facebook.com/SustainableSmallholding/

Twitter Lorraine Turnbull
https://twitter.com/LorraineAuthor

Instagram Lorraine Turnbull_
https://www.instagram.com/lorraineauthor/

Acknowledgements

Thank you to the many people who contacted me to suggest I write this book following the success of my first book; *The Sustainable Smallholders' Handbook.*

Special thanks to Stuart Bache, bookscovered.co.uk, Tina & the late Richard Holland, David Esnaut & Delphine Remion, Jean-François Confaveux, Pierre Joinel, Elspeth Prins, and to Valerie Outfin, Rod Haselden-Nicholls, Joanne Ainscough & Cheryl Arvidson-Keating for reading the book and suggesting useful changes and layout. Finally, huge thanks to my family for encouraging and supporting me as always.

About the author

Lorraine Turnbull wanted to be a farmer since she was five years old. After running a successful gardening business in Glasgow she uprooted herself and her family and moved to a run-down bungalow with an acre of land and an Agricultural Occupancy Condition in Cornwall. She retrained as a Further Education teacher and taught horticulture at both adult & further education level, whilst running a one acre smallholding. After working as a Skills Co-ordinator for The Rural Business School, she began commercial cider making in 2010.

In 2014 she was recognised for her contribution to sustainable living by winning the Cornwall Sustainability Awards Best Individual category. She successfully removed the Agricultural Occupancy Condition from her home before moving to France. Her first book, *The Sustainable Smallholders' Handbook* was published in May 2019.

Lorraine now lives in the Dordogne with her husband, cocker spaniels and sheep. She is currently writing her third book, on UK Agricultural Occupancy Conditions.

Contents

INTRODUCTION

Fed up with the rat race? Dreaming of a simpler life? A better life? A GOOD LIFE?

Since the 1990's hundreds of thousands of people left the UK behind and moved to France.

Are *you* dreaming about moving to France to perhaps live happily ever after on a smallholding or simply in the countryside? Then THIS is the book for YOU.

This practical and up-to-date book will lead you through the many questions you may have including:

* How Brexit may affect you

* Owning animals or setting up a smallholding in France

* Finding & securing the right property

* Starting a rural business

Benefit from Lorraine Turnbull's own experience and read the case studies from real people who have moved to live the Good Life in various areas of France. It's a big step to a brave new world and this timely book aims to help you in your journey to your Good Life in France. Lorraine is an award winning smallholder and former commercial cider maker who relocated to rural France in 2017.

Chapter 1
Your French Rural Dream

So you're thinking of moving to France and living the 'Good Life'?

Well, I'm delighted that you've decided to read this book before making this colossal move, and I'm sure you will find plenty in it to inspire and help you achieve your dream.

The dream is of course that you'll be able to afford a nice property, possibly with a bit of land; that the weather will be sunnier, warmer, and that life will merely seem like an extended holiday in the sun. Those Peter Mayle books you have read and re-read paint an idyllic picture of lazy French life and YOU want your share!

This book is aimed at those with a dream of moving to France; but is especially aimed at those wishing to live in the countryside; perhaps owning animals, growing their own produce and living a slower pace of life. You may have been considering smallholding, which I can thoroughly recommend if you want to keep a few animals, know where and how your food is produced, care about being even partly self sufficient and want to live more in tune with the land and the seasons. Of course, this lifestyle isn't everyone's dream, and this book is also suitable for those who just want to relax into the French lifestyle, weather and great food and wine. This book is full of information and includes a few easy recipes (I'm NOT a cook) to introduce you to great French produce the French way.

So, what will this book do for you?

Well, it will help you ask yourself some questions which will help you decide *where* you want to be in France, *what* you want to do, and *how* you can get there. It will also explain generally about the bureaucracy involved in buying a home, basic business information (if that's your aim) and about the requirements to live 'in the system'. It also explains a little about what to expect from rural life in France.

The reality is that Europe is undergoing some big political and climatic changes right now. As I write this the UK has just left the European Union and the future; especially after the one year transition period is very uncertain. This naturally makes all Europeans think about stability, the price and nature of everyday life and the future. For some people this is unsettling and many will want to consolidate what they already have; but for others this is a time of opportunity and change and many see this as the time to make that lifestyle change they have been dreaming of for so long. From the 1st February 2020 Britons in France will no longer be considered to be EU citizens; however in terms of practical rights and procedures, nothing much will change until the end of the Transition Period in December 2020. It is expected that after the Transition Period Britons moving permanently to France will require a long stay visa, but will still be able to buy property in France the same as they did before the UK joined the EU in 1973. I will try and comment on the known changes where relevant in this book.

Climate change also significantly affects our lives. Those 'once in a lifetime' floods, storms and droughts are becoming ever more frequent and we look around us to see if moving elsewhere will lessen the effects. These changes are happening all over the world and awareness of making even small changes to the way we live, work and eat can only be a good thing.

Whilst many in the UK and in the USA think of a more sustainable way of life 'at home'; many are thinking of a complete change and considering a self sufficient way of life on the continent, and France is an increasingly popular destination. Whilst *La Belle Vie* is an attractive option you must embark on such a massive move with your eyes open and this book aims to give you a lot of information to help you decide if this is a sensible choice for you.

The attractions of rural living have appealed since before the 1960's and whilst 'living the rural dream' is an aspiration many have in the UK and USA, transporting this dream to France and making it a reality is not a walk in the park. Having said this, there are many happy individuals of all shapes, sizes and ages in France enjoying living in their rural or village cottage or their smallholding.

Yes, the language difficulties and France's love of a paper trail are difficulties, but these are not insurmountable, and knowing what information is required and where to find it should ease most people into moving to their own rural heaven. Information is indeed power.

If you have never lived in a rural area in either the UK or USA, you are going to find rural life in France a shock to the system. Most supermarkets and shops have early closing (here in rural South Dordogne it is 7.30pm) and closed on Sundays from 12.30pm. Small communities are close knit and you will really have to make an effort to join in and integrate and of course this means having some command of the language. In our village, streetlights are out at 10pm and most people are up with the lark. The church bells ring at 7am, noon and 7pm, telling you when to get up, when to eat and when to stop work. August is the big holiday month and even government departments are understaffed and everything takes forever to get done. Like most nations, France has its idiosyncrasies, which some find annoying and some endearing.

In winter the population shrinks in the countryside. Those with holiday homes return home; whether that is to the UK, USA or Paris. The elderly batten down the hatches and hibernate until spring and you only see your neighbours on supermarket visits, unless you make a real effort to get to know your French friends and perhaps invite them round for lunch or dinner.

Winter in rural France can be the ultimate test of your determination to live in France year round, and can put strain on the most stable of relationships, your mental health and attitude to loneliness. It is quite unlike living rurally in most of the UK, where civilisation is a short drive away. Just this year we have heard of two couples who have left the small community here and the winter isolation had a part to play in both decisions.

If you dream of having some chickens or a few sheep or goats on your French property be aware that just like in the UK, you must comply with French law as regards keeping livestock, and also face the prospect of pests, diseases and death. This isn't as bad as it sounds because in the UK we have been adhering to European Union laws and regulations for many years and most of the rules in France are the same. Keeping livestock however, is a 24/7 lifestyle and arranging animal cover for holidays or trips back to the UK or elsewhere is either problematic or expensive. (Now, *there* is a business opportunity for someone enterprising – see my chapter on how to make it pay).

This book aims to give you the answers to many but not all questions you may have about the enormity of starting a more rural lifestyle in France. Of course there are some who will want to fully embrace the 'Good Life' more than others, and some who just want to live a quiet rural life in the countryside with no animals. There is information about starting a vegetable garden or *potager*, about orchards, and about managing woodlands. A few recipes are included; after all part of the joy of living in France is to embrace the art of cooking and preserving your produce.

When we first moved to our smallholding in Cornwall, we thought an acre was a huge amount of land and envisioned only having poultry and bees. Within 3 years, we realised how small an acre is when you want to have a breeding flock of sheep and a few ducks and an orchard. When we moved here to France the idea was to 'retire' and animals were just not going to happen, and yet we annually have lambs for fattening and our one acre of woods isn't as big as we would have liked. However; at our stage in life we are reigning ourselves in. There will come a time when the land we have will become too much for us and the house too large to maintain. Until then, we take each day as it comes and count our blessings.

I don't propose to include the translations for all the words or phrases in the text; after all, part of the fun of moving to another culture is the learning of a new language; so there is a basic glossary of terms in English and French at the end of the book and lots of helpful websites and addresses to further research for information.

Moving to France is possibly the biggest adventure you will ever consider; yet every year many take a deep breath and do it. British people 'back home' think us 'terribly brave', although British people already living here wonder why we took so long! There are many motivators for such a massive life changing decision and they are as individual as the people making that decision. And, as one of the prime reasons people move to France is the better weather, what better way to start this book than with that old British conversational favourite – the Climate?

CLIMATE

France is a *big* country; approximately twice the size of the UK with a similar population. This means that outside the big cities there are a lot fewer people per square kilometre. Such a big country enjoys four distinct climatic zones, although generally France enjoys a temperate climate, with the divide being roughly the Loire river valley. North of this is cooler and wetter, south of this is hotter and drier. Of course, this is a generalisation and, like any other location in the world there will be local microclimates, which is why you need eventually to be searching for your ideal location on the ground.

In Normandy, Brittany and down to the Loire Valley the climate is very similar to that of the UK and enjoys an Oceanic climate with average rainfall with a climate similar to south western UK. Central and Eastern France have a continental climate with cold winters and hot summers (Champagne, Burgundy and the Alsace regions). South East France (Provence, Cote d'Azur and Corsica) have hot, dry summers and rain from October to April; The West and South West (roughly the Aquitaine region) have mild winters and hot summers, and finally at altitudes of over 600 metres, a mountain climate is the norm, with heavy rainfall and snow from three to six months of the year.

Naturally, the climate will be one factor governing choice when you are looking at areas of this beautiful country to start your new life, but bear in mind that climate will also govern what you can do on a smallholding. For instance, if the hot sun burns off all your pasture in mid summer, you will still need to provide food for any livestock you plan on keeping. This may significantly increase your running costs on a smallholding. Also, property prices may be much higher in areas with a sought-after climate, and lower in other areas.

You may have other reasons for your choice of region. If you love skiing and winter sports then the Eastern or Southern mountain areas may be high on your list. If you want access to the sea then the West or South East of France may be high on your tick list. If you are very fair skinned then the hotter areas may be just too much for you on a permanent basis; and of course those moving with career may need close access to cities or fast transport links.

If you aim to have a self-sufficient lifestyle your chosen location will need to produce an abundance of fruit, vegetables and other products. Just like in the UK there are many areas with local microclimates, but there is no substitute for visiting such areas to assure yourself that these areas fit the bill and can provide all you require.

The cost of utilities in France is also higher than in the UK, so bear this in mind when considering heating costs. Wood burners (*poêles à bois*) are common in rural properties and are wonderful to enjoy in cold, sharp winter days, but a real pain and expensive to keep running all day and night from October through until April. And ask yourself where is this wood fuel coming from? Are you prepared to either buy in your firewood or to harvest, cut, dry and stack your own in summer and autumn for use two or three years down the line? Electricity is expensive in France compared to the UK and mains gas supplies infrequent in rural areas, although gas cylinders are common for cooking use in rural areas. Although government support for renewable energy is talked about often, the uptake is slow due to high costs, but hopefully this will change as necessity and climate change demands.

Water provision and use is increasingly important. Whether you have your own supply or have water from your town, the availability and use of water is becoming a significant worry throughout France. If you have livestock, potagers or gite guests, water will be an Issue for you to think about, and plan for provision. Building a *citerne* (water storage tank) is a huge expense, but may be a requirement in the future.

Internet provision, reliable wifi and mobile phone coverage may be a necessity to you, your business or gite guests. Despite what providers tell you, you will need to examine the coverage in any area you plan to live in.

Also, remember you will not be on holiday. Having a fortnight of blazing sunshine is fine on the beach, but watching it shrivel your prized crops is another thing altogether. Is the area you are looking at going to require air conditioning in the summer? Are you prepared to *pay* for the installation and high running costs of air conditioning?

Freak weather is on the increase also, with hailstones the size of snowballs common during summer thunderstorms (*l'orage*) in the Charente causing damage to crops, houses and vehicles, and high winds possible all over France. Flooding is on the increase, with many larger rivers bursting their banks every spring with rainfall and melt water from the Alps. Here in Dordogne, the weather can change rapidly with frequent late afternoon thunderstorms bringing huge amounts of rainfall in a few minutes and thunder and lightning like you've only experienced watching old Hammer horror films. For all the above reasons, be cautious about buying property on the flat alluvial plains of ancient rivers, in for example, the Gers or Hautes Pyrénées, which can be prone to flooding, and if not, very boggy fields.

One of the best sites to monitor the weather is **www. meteofrance.com** and you can get vigilance warnings sent to your mobile phone to warn of storms (l'*orages*).

FINANCIAL CONSIDERATIONS

It's a widely held notion that the cost of living in France is lower than that of living in the UK. Let's just break this down and look at it. Yes, the price of some older properties in certain parts of France can indeed be a lot cheaper to buy than a similar property in the UK. Younger French people like modern houses. They like the convenience of central heating, good insulation, and mod cons, and they need to live near to where they can find work. Consequently, the old family house owned by parents or grandparents is seldom attractive; especially if it's located off the beaten track or far away from work in the city. Thus in France older country houses tend to be cheaper, which is great for the non-French buyer who is looking for a character house with a little land. However, renovation of an old house is expensive in France. Whatever you think it will cost – double it!

Selling your own house in the UK, paying off the mortgage and moving out with a huge pot of money seems like an ideal situation, but unless you do a lot of soul searching and research, perhaps you would be as well to rent; especially through a winter, just to see if your expectations are indeed being met. Surely it's better to take your time and get the right place first time than make an expensive mistake.

Look at the internet, newspapers, specialist agencies and estate agents and also inquire at the local *mairie* (town hall), who may know of a property that may be suitable as a long term let of at least 6 months; but avoid the summer holiday month of August and the *rentrée* (return to school) months of September and October. In the popular University cities students are still chasing accommodation at this time. Common sense dictates that you search to rent in the area you are considering buying in, and don't expect to be able to rent a smallholding; this is hard enough to do in the UK, and nearly impossible in France. A couple of national agencies offering long term letting are Seloger http://www.seloger.com and Rent a Place in France http://www.rentaplaceinfrance. com although there are others.

Once you have found yourself a long term letting property you can begin the exciting and tiring process of finding a permanent home. If you decide in the rental months that you have second thoughts then it's a less dramatic mistake than having already purchased a property. I'd not look at this as a failure – rather as a planned trial that allowed you to experience living in France before committing either way. It is worth mentioning that house prices are stable or falling, like the potential for the exchange rate unlike in the UK, so going back may be very difficult.

A cursory glance at the online French estate agents will reveal some beautiful homes at apparently affordable prices. BEWARE! Photographs can be misleading. The unsuspecting buyer is unaware how old the photographs are, of what lies beyond the tree line and what the property next door is. It may be a nuclear power station, a pig farm or a disused industrial unit; who knows? Read the estate agent's description of the property with a pinch of salt and use online maps and gather as much information on the area to give you an informed knowledge of the area. Another annoying practice is to list the property as 'near to' a certain named town. This is to prevent the buying and selling of homes outside the control of the estate agent who then would not receive his (large) commission; and 'near to' could actually mean anything between a kilometre and 30 kilometres.

There is also a unique system of house selling in France,

called the *viager* system where a typically older home owner offers a house for sale, but continues to live in it until his or her death. This is a high risk purchase for the buyer and really you MUST take professional advice if you wish to consider this: French people can live a *long* life, and you will also be responsible for paying a lump sum (known as a *bouquet*), as well as monthly payments to the seller for the rest of their life. The buyer is also responsible for any major works required to the property and for land tax.

In practice, it is similar to the equity release schemes currently becoming popular in the UK, although the *viager* purchase is a private transaction, overseen by a *notaire* or solicitor.

If the seller is typically aged around 70, for example, the property value is calculated around 50%. The *bouquet* (lump sum) would be around 30% with the rest in monthly instalments. If the buyer defaults on the monthly payments, the seller keeps the *bouquet*, the monthly instalments made to date and retains ownership. For the buyer, the system offers a chance to purchase property cheaply, a reduced *notaire*'s fee and is suitable for experienced, high risk investment.

If you intend to take out a French mortgage please do your homework and take advice from a specialist in French law. Mortgages are currently available to European Union citizens who fulfil the necessary criteria and your bank or mortgage provider will happily provide you with a pre-approval if you meet their requirements. Certainly being a cash buyer or having a mortgage will in principal will give you the upper hand in negotiations when purchasing, as many French houses are on the market with multiple *immobilières* for years.

Once you have set a budget for the house itself, you need to factor in *notaire's* fees. These take a hefty chunk of cash (between 2 and 8% depending on the property price) over and above the cost of the house; part of which goes to the French government (the French equivalent of Stamp Duty) and part towards the conveyancing. Who is paying the estate agent - you or the vendor? If the vendor then look for the words *frais charge de vendeur*, and if the buyer is paying then the price should show this and look for the words *frais charge acquéreur*. Does the property schedule or internet advert have the letters FAI after the price? If so then this means the agents fees are included, but NOT the *notaire's*. Be wise and check these details in order to avoid a nasty surprise or worst still, finding that perfect property and then realising you just can't stretch to the price.

Ensure you are aware of any works that require to be done now or in the future. How good is the roof? Does the septic tank (*fosse septique*) comply with the latest regulations? Replacing a fosse can cost between €8000 and €12000 depending on the difficulty of the job/terrain etc. Is there any damp? Does the property need rewiring? Ask yourself the usual questions you would ask when looking at a property in the UK. In addition, find out if there are any rights of way, easements or traditional agreements with neighbours regarding land use etc. Ask what the situation is regarding the family relations and the sale of the property. This may seem odd, but its amazing how many property purchases stall or even fail because suddenly the family refuse to agree to sell; especially if the house has had to be marketed due to an inheritance amongst children.

Is the property already fenced? Fencing can be extremely expensive to install and then to replace when required, but its essential if you intend to keep animals in or out. Animal housing too – goats and donkeys really do need waterproof shelters or outbuildings, so are there any?

Moving on from properties; you need to live. France does not have a National Health Service so you need to ensure

22

you have provided for medical cover. Whilst the UK was a member of the European Union, and until the end of the Transition Period in December 2020, a retired Briton living in France could take advantage of the S1 arrangement; where people of state pension age have free healthcare in France. The Withdrawal Agreement with the EU means that Britons living legally and permanently in France will be able to stay and continue to work and access education and healthcare. Those living in France for five or more years at the end of the Transition Period are covered by the Withdrawal Agreement and entitled to the rights agreed to by the EU and the UK. If you have been legally resident for less than five years by the end of the Transition Period, you will be allowed to stay and make up these five years and meantime be covered by the terms of the Withdrawal Agreement. You should be able to access healthcare as before, via your carte Vitale and if you are a UK state pension recipient via your S1 arrangements. Please make enquiries via the Department of Works and Pensions in the UK for up to date information.

If you have moved to France and are either an early retiree or of working age and wish to take advantage of France's excellent healthcare, you will either have to pay for private health insurance or join the system by becoming employed or (more likely) self employed. Being 'in the system' entitles you to a *carte Vitale* (Vital card) which you show to your health professional when you need treatment. If you also have top up (paid) insurance via a *mutuelle* you will be covered for most eventualities.

You will have to budget for expenses such as running a vehicle, electricity, telephone, water, property tax, house insurance, the cost of buying livestock, fencing, housing, vets bills, animal food bills, and a hefty lump sum for emergencies. The removal itself will have to be budgeted for. Even for a mature couple with no children you will have personal belongings and probably a vehicle to bring over. People often ask about whether to bring their existing furniture; this is a difficult one. We got rid of most of our large Items in the UK, but I have regrets and haven't found good quality replacements for the kind of price I'd be willing to pay, but friends have told me to save money on the removals cost and buy French. Then there is the family dog, cat, horse, and any useful tools or machinery. My 'retired' husband insisted we brought two welders and his blacksmithing equipment; whilst I brought all my commercial cider making equipment.

Most removal firms will price per cubic metre. Ensure any firm you choose has adequate insurance and whether your agreement is for a part load or full load. There are many English speaking firms in France who offer reasonable price removals in both directions. Ask around for recommendations – social media groups in France are an excellent source of information on a range of subjects and in different parts of France; and expect a large price for a decent job.

Finally, you can start to budget for food and drink. Whilst wine is naturally cheap in France, beer is not. Meat is slightly more expensive in France than it is in the UK. Remember food is a sensory, almost religious experience here, whereas sadly in parts of the UK and the USA food is seen primarily as fuel. Yes, as a smallholder or rural dweller you may be going to produce a lot of your own food, but it's best to budget for buying your own, at least in the beginning. Home produced lamb takes roughly five months from birth to butchering, and even humble fruit and vegetables need time to establish and favourable weather.

CHILDREN AND PARENTS

Making a lifestyle change does not just affect you. If you are married or have a partner, you must ensure that you are both equally keen to embrace not only living in a different country, but also living in a possibly more isolated area. You will have to make new friends, learn a new language, and if you are intending to run a smallholding there are many new skills to learn. If your partner hates flies, muck and the odd furry, feathered or even slithery creature taking up residence in the house then perhaps you need to re-evaluate your choices; whether that re-evaluation is of your lifestyle choice or of your partner!

Many marriages and partnerships have failed within a year or so of moving because one or both of the partnership have underestimated how big a change they face. They flounder with losing the network of friends and work colleagues from back home and sometimes have to spend all day with a partner who they previously only saw in the evening or weekend. Add into the mix the pressure of learning a new language, and dealing with the inevitable paperwork involved in establishing yourself in a different country and you could have tensions rising.

If you have yet to reach middle age you may be bringing children with you as part of your move. This means new schools, new friends, a new educational system in a new language, and can also mean losing friends, being lonely or depressed. If your children are young they will adapt quicker, but I would caution moving teenagers. They are hard enough to handle without the added difficulties of new language, social acceptance into established peer groups and possible educational setbacks. Add to this the startling idea that mum and dad are going to be "yoghurt knitters" or farmers and you have a recipe for disaster.

There are some private international schools in the larger cities, but if you are looking to run a smallholding or live the rural life you may find that unless you are prepared to place your children into a private international boarding school

25

this is not the solution for you.

You must also consider the family members you are leaving behind. This usually means mums and dads, but could also mean older teenagers. Now mum and dad may have been enthusiastic and supportive when you first came up with the idea of moving to the French countryside; but as the reality starts to dawn, many will realise that they won't have you or the grandchildren visiting as often. They may worry about the future in the event of either of them being ill or dying and the survivor having to cope alone with no family nearby.

Naturally, you may worry about them too. You may even think of suggesting that they come out and live with you! Time to put the brakes on and put that kettle on folks!

Do you *really* want to have elderly family living with you? Be honest now. For a start this means a bigger property. Then there are the additional healthcare costs. Will they settle? Will they learn the language? Are you going to have to take on the job of Entertainment Officer for them too? What happens if they have a stroke/ heart attack etc? Do they have the financial independence required and more importantly, can they prove it? And of paramount importance will you all get on? The time to have this important and honest discussion with your partner is earlier rather than later.

LEARNING THE LANGUAGE

If you are serious about moving to France start to learn the language as soon as possible. Although certain villages have a huge non-French population and English is widely spoken, you will gain more respect and get more out of life in France if you can speak even just a little of the language. If you end up only with English speaking friends ask yourself if you would have chosen the same people as friends back in your own country. Besides; you chose to move to France with all its delights; and learning a new language will ensure you get the utmost from the experience. You are never too old to learn and acquiring a new language keeps your brain exercised. Don't take yourself too seriously; and even if you don't always get it right, you have tried, can laugh about it and try again.

There are many community based classes in the UK run at very little cost and you can begin to learn at a pace that suits you. Have a look at Adult Education classes run by your local council. Try looking at French newspapers online to get a flavour of what is going on in France and the attitudes on certain subjects. There are even apps you can get for your mobile phone with daily French bite-sized lessons, such as *Babbel*. My husband and I like learning from the BBC language website **www.bbc.co.uk/languages/french**.

When you do arrive, continue to learn. Some local towns have a sufficient non-French population to warrant French language classes. Some of these are run by the ex-pat community, some offered by the local *mairie* (town hall) to help integrate English speakers into the language. You can increase your own knowledge by taking private French lessons too.

Being able to converse with your neighbours and make local friends is invaluable. Remember, they will be just as curious as you are and want to know why you chose their area to live and what you intend to do. In a small and close knit village the population will want to know if you are going to be summer visitors or permanent. Many villages in tourist areas are empty in winter, which is not good for local businesses and breeds resentment. But if you intend to be permanent and active members of the community you will be welcomed with open arms.

Regulars at the *bar/tabac* will always acknowledge you with a *bonjour,* and as they get to know you this will turn into a handshake or the Gallic double kiss if you are female. The tentative friendship will only grow if you can manage to order your drinks or remark on the weather in French.

Watching French TV, especially adverts and listening to French radio is a good way to immerse yourself in the language and you will soon start to pick up individual words and then phrases. If you play the radio in the car, you will start to absorb the language subconsciously. Joining local groups such as walking, gardening or history groups will help you integrate, make new friends and immerse you in the language. There is always someone who speaks some English and it's a good way to become more involved locally.

Of course you need to keep up with events back home and can also manage to receive UK television in various ways. It's nice to keep up with the latest must-see costume drama or favourite soap, especially in winter. And if you want to see the latest movie release, many French cinemas, especially in larger towns occasionally show films in their original language *(version original* or VO).

INTEGRATING INTO FRENCH LIFE

When you finally buy or rent a property and have moved into your chosen home you will naturally want to go 'meet the neighbours'. Perhaps you will be lucky enough to be invited round for *l'apéritif,* which is for drinks in the early evening. This is an honour and although may make you anxious, it is an opportunity to meet your neighbours. They may already know who you are from the sellers or from the local grapevine; especially if you are foreigners and they will be keen to assess you. Taking a gift of a bouquet of flowers (*not* chrysanthemums - they are funereal flowers) or box of dark chocolates (milk chocolate is for children) will be very welcome, but taking a bottle of wine would be a *faux pas* (a mistake); unless it's a bottle of Champagne, which is always welcome. You'll not get a drink until everyone has arrived and you need to wait until your host has toasted everyone before you take a sip. If you are unsure how long to stay, your host will give you the subtle hint to go home when he stops refilling your glass.

If you are invited to dinner or lunch be prepared to follow what your hosts do and try to do likewise. This includes rituals about bread – placing bread on the table or on your napkin near you is more acceptable than on your plate as is having the baguette the right way up; and also if soup is offered, some may follow an old French custom of tipping a little red wine into the last of the soup in their bowl and picking the plate up and drinking it! Be careful with this! In fact there is a whole etiquette about what to do 'at table', so best read up on this online before you go to dinner.

The French; especially those living in *la France profonde* (the French countryside) are extremely polite and welcoming in general, although sadly there are surly neighbours to be found everywhere.

Remember when talking to someone new to always use the formal *'vous'* until you are invited to use the familiar *'tu/ toi'*. This is especially important when talking to the older generation, although children are always addressed in the familiar, and expect to be kissed on greeting. You shouldn't attempt to kiss adults when first introduced, but the *faire la bise*; whether one, two or even three kisses to the cheeks will be expected when you have become familiar. Here in our village, it took almost a year before we were on *'tu'* and kissing terms with our French friends and neighbours.

My husband is very quiet and one of our elderly neighbours

is very deaf and complained to another neighbour that my husband was rude and made no effort even to reply to his *bonjour*. The situation resolved itself pleasantly one afternoon as I chatted to the old man's son who was taking an interest in my sheep. I explained in my poor French that my husband was quite quiet and was mortified that our neighbour had taken offence at his mumbled greeting. The son explained to the father and the gift of some lamb chops from my husband at butchery time meant the incident was a thing of the past.

A visit to your local *mairie* to introduce yourself and also register your new ownership of a property to ensure the water supply is a good opportunity to ask any questions, find out who the local plumber and doctor are, etc. You may be alerted to local customs such as the practice of observing quietness between 1200 and 1430 daily and after 1200 on a Sunday. The French long lunch and Sunday family day are times when food and relaxation are enjoyed. Running any machinery, power tools or lawnmowers at these times will not endear you to your neighbours.

It is also now illegal in many (but not all) parts of France to have a bonfire at all on your land, unless you are a farmer, and I would ask at your local *mairie* for guidance on this matter. In our commune we can complete a form and get prior permission to have a controlled and supervised fire between 1st October and 28th February each year. There will always be those who ignore such laws, but as a newcomer, err on the side of caution. Garden waste and large items such as broken furniture or white goods can be disposed of at your local *déchetterie* (rubbish tip). Some *déchetterie* charge or use a card system to regulate use, but some, like ours, are small and friendly and as long as we are clean and respectful when getting rid of our waste, they are equally friendly to us.

During summer and autumn, you cannot fail to notice an

abundance of fruit and nuts just sitting there at the roadside or on apparently abandoned pieces of land. Please be aware that these do belong to someone, and it is considered theft to take from the tree. If fruit or walnuts, for example fall onto the public road, then you may collect, but to be honest, if you approach friends or neighbours who have plenty, they will probably be only too pleased to let you pick some.

In the run-up to Christmas, you will suddenly be visited by the *Pompiers* (firemen), who offer you their wonderful calendar in exchange for a donation. This will usually be followed by the postman or woman, the local football club or school, who will all offer you calendars. A donation of between five and ten euros seems to be the acceptable amount.

Most villages have local events, such as *vide-grenier* (car boot sales), *fête votive* or similar and if you are lucky enough to have a thriving village with a school there may also be a walking group, history group or similar. Parents with primary school aged children have an opportunity to get involved with the French equivalent of the PTA. Also meeting your children's friends is a good way to get to know people locally.

If you live in a small village or community and there is a death within the population, you may only know through gossip, the slow toll of your commune's church bell or a notice pinned on the church door or village notice board. Take heed of the date of the funeral and make the effort to attend if possible. French funerals are short affairs, but if the deceased is a neighbour it is considered bad form not to pay your respects. Funerals take place very quickly after death and locals attending may appear more casually dressed than they do in the UK or USA.

If all of this sounds very daunting and you are frustrated at the different system together with the language difficulties,

31

be aware that there are many hand-holding services available. Katey Warwick offers such a service and works remotely to cover the whole of France. She offers help with schooling, healthcare, benefits, *Carte de Séjour*, and importantly with registration as a *Microentrepreneur*, and in the *Maison des Artistes* and *Chambre d'Agriculture*. So contacting her initially via Facebook might offer a way forward if you know what you want to do. See the appendices at the rear of this book for contact details. Karen Jones offers another similar service, offering help with business set up, and accounting for the French system. Karen's details can also be found in the appendix.

You don't have to be sporty to enjoy wandering around the many local produce markets or to enjoy going to a wine tasting or *dégustation*. Many areas of France are wine producing areas and organise tours of the vineyards and *dégustations*. Champagne, Burgundy and Bordeaux are probably the better known ones, but even near our village we have a small producer prepared to sell at the gate following a tasting.

You can combine a trip to beautiful landscapes filled with vineyards with an educational visit about the *terroir*, the different varieties and the production process and finish with tasting three or four wines. This allows you to stock up with one or two bottles or one or two cases. Wine tasting at the cellar is usually free, but you are expected to buy at least a single bottle.

Beware if you are driving, and try and limit your tasting to three or four different wines. Experts cleanse the palate between wines with a little fresh baguette or spit the wine into a bucket provided.

Harvest time is September/October and the wine producers are busy, so best to avoid or book in advance. Autumn is also the time when supermarkets tend to hold 'wine fairs' to encourage bulk buying.

DEALING WITH AN EMERGENCY

Dial 112 ANY emergency from a landline or mobile phone

Dial 15 Ambulance & medical emergency (SAMU)

Dial 17 Police

Dial 18 Fire Brigade (sapeur-pompiers)

The SAMU deal with very serious emergencies and often the *sapeurs pompiers* initially attend road accidents and medical emergencies.

Key phrases translated

Police: *La Police Nationale or gendarmerie*

Fire brigade: *Les sapeurs pompiers*

Emergency services/ambulance: SAMU

Poisoning emergency: *Urgence d'empoisonnement*

Road emergency services: *Services d'urgence routière*

It's an emergency: *C'est un cas d'urgence.*

My name is…: *Je m'appelle…*

My telephone number is…: *Mon numéro de téléphone est…*

I live at…: *J'habite à…*

Help!: *Au secours!*

Ambulance: *une ambulance; J'ai besoin d'une ambulance.* (I need an ambulance.)

Heart attack: *une crise cardiaque: Mon mari fait une crise cardiaque.* (My husband had a heart attack.)

Stroke: *une attaque cérébrale; Je pense que ma femme a souffert une attaque cérébrale.* (I think my wife suffered a stroke.)

Choke: *s'étouffer; Mon bébé s'étouffe.* (My baby is choking.)

Difficulty breathing/gasping: *haleter* or *difficulté à respirer; J'ai difficulté à respirer.* (I have difficulty breathing.)

To bleed: *saigner; Je saigne beaucoup.* (I am bleeding a lot.)

Hemorrhage: *une hémorragie; Mon mari fait une hémorragie.* (My husband had a hemorrhage.)

Concussion: *une commotion cérébrale; Mon enfant est tombé. A-t-il une commotion cérébrale?* (My child fell. Does he have a concussion?)

Diabetic: *diabétique; Je suis diabétique. J'ai besoin d'insuline.* (I need insulin).

Labour: *accouchement/accoucher; Ma femme accouche; la poche des eaux a percé.* (My wife is giving birth. Her water has broken.)

To be poisoned: *s'empoisonner; Mon enfant s'est empoisonné.* (My child has been poisoned.)

Of course, you will be keen to continue with any hobbies you had before moving to France, or indeed to take up new ones. The climate, on the whole is very conducive to outdoor sports and activities.

Walking or hiking (*tourisme pédestre*) is a national sport and taken seriously enough for a huge network of marked paths and trails. The more popular tourist areas include the Alps, Pyrenees, Vezere Valley, and the Auvergne, and also the national parks and around the coast. The *Grand Randonée* network (GR) is a huge network of footpaths. Remember, especially out of season the footpaths and trails can be very quiet. Please let someone know where you are going and when you expect to return. *Michelin* orange series maps are pretty detailed, but IGN *Rando* series maps provide more detail.

Cycling in France is taken very seriously, with national coverage of the Tour de France and other cycling events. Bicycles can be bought from supermarkets and from specialist shops and there are many cycling clubs all over the country. Mountain biking (VTT) is popular and there are many trails and clubs; details here: **www.vttfrance. com**. There are useful websites aimed at tourists but equally useful for keen resident cyclists to make use of such as **www. francevelotourisme.com** and **www.freewheelingfrance. com**. Of course, cyclist safety is paramount and cyclists are advised to wear bright clothing, helmets and ensure that lights and reflectors are all working and clean.

Fishing is another popular pastime at the sea and also inland on rivers, lakes and ponds. Assume that any inland water is protected, and that you require a permit or license. More information can be found here: **www.cartedepeche.fr**

Golf is one of the fastest growing sports in France, with many clubs in North West France and around Paris and the Côte d'Azur areas. The Fédération française de golf is a good place to find information on course locations **www. ffgolf.org** and also **www.touslesgolfs.com**

And of course, there is skiing. From Christmas through until Easter in some years, skiing is available at all levels and for all ages in the mountain ranges in France and some resorts even offer summer skiing, although with global warming this is rare now. Most ski resorts will have a Ski school for beginners of all ages - ESF (Ecole de Ski Francaise). They also have a good understanding of the mountains locally. Downhill skiing is more popular than cross country skiing, but other winter sports are also available at resorts including skating, tobogganing, curling, dog-sledding, snowmobiling and indoor sports and the usual après ski activities. Be aware that there are no voluntary Mountain Rescue Teams and rescues are carried out by the PGHM (Peleton des Gendarmes d'Haute Montagne) and may involve helicopter transport. This can work out very expensive and you should ensure that you have appropriate insurance.

CAF (Club Alpin Francais) is the national climbing club (lots on FB) can include insurance in its membership fee and probably the best place to start with enquiries about mountain sports in general.

Tourain blanchi (Soupe a l'ail de Sud-Ouest) Garlic Soup

1tbsp duck or goose fat
8 or 9 cloves of garlic
4 sage leaves (optional)
2 heaped tbsps of flour (farine de ble)
1.5 litres cold water
2 eggs
1 tbsp good wine vinegar
Salt & pepper
Some slices of pain de campagne (country bread)

Melt the fat on a low heat in a large deep pot. Peel and crush the garlic and add them (along with the sage if wanted) to the fat and cook slowly until golden or translucent. Remove and discard the sage leaves. Add the flour and mix. Slowly add the cold water and the salt and pepper and stir to mix. Boil for 20 minutes and then simmer on a very low heat.

Split the eggs and save the yolks for later. Beat the whites into the soup, stirring a little. Simmer a further 15 minutes.

Put the egg yolks into a bowl, add the vinegar. Remove the soup from the heat and allow to cool a little before adding the yolk/vinegar mix, beating it in. At this point the soup will immediately 'rise' alarmingly in the pot like boiling milk!

Serve by placing 2 slices of country bread into the bottom of soup bowls and pouring the soup onto the bread.

Chapter 2
Finding, Buying & Moving To France

LOCATION, LOCATION, LOCATION

Spending considerable time deciding where you want your permanent home in France is time well spent. France is a big country with a lot of variety, so don't assume that you will be able to spend a week visiting and magically find the perfect property right away. You are about to make a huge financial investment in a country that is a little different from the one you come from, away from family, friends and the old life you have been accustomed to; so ensure you are taking enough time to consider fully all the factors you need to before making a decision.

Deciding whether your house in France will be a main residence or a second home will vastly affect your tax situation. I will just briefly mention the Capital Gains Tax situation as an example of why consulting with an expert in French law and taxation is money well spent.

You are considered to be fiscally resident in France (therefore liable for tax purposes) if you live in France for 183 days or more; If/whether you are of retirement age or carry on a professional activity (whether self employed or employed) and you have a centre of economic interest in France. The tax year runs from January 1st to December 31st.

Capital Gains Tax (CGT) is liable on the profit made from the sale of a property and in particular (but not exclusively) from the sale of a second home in France. If you are resident in France and selling your MAIN residence you are exempt from CGT (*impôt sur les plus-values*). However, if you are selling a second home you are liable to pay CGT unless you have owned that home for thirty or more years. The amount liable for CGT is calculated by deducting the purchase price (and any costs for renovation and improvement) from the sale price. To prove the cost of renovations or improvements you must produce proper invoices with Siret numbers. The *notaire* will calculate the CGT due and deduct this before passing the sale proceeds to the vendor.

The most important thing is to choose the right area. If you wish to live a rural life with poultry or animals then a city will not be the right choice, but most large towns or villages have plenty to offer in the way of property and services. Getting back to the UK or elsewhere quickly may be a huge bonus, especially if you intend to run a hospitality business from your property. Holidaymakers prefer to be within an hour's drive to an airport connected to the UK or within a few hours drive of the ferry ports. If you have medical concerns then access to a large hospital may be a factor influencing your choice of area.

If you intend growing crops or fruit or keeping livestock, the area and the associated climate will also be important. Scorching hot summers and freezing winters are to be avoided. If you wish to sell your produce then proximity to a nearby town or city with lots of markets might also be a consideration.

Do you want to live in or near a town with a large ex-pat

/immigrant British community or live a more complete French lifestyle? If you are a vegetarian or vegan will you feel at home with the French pastimes of hunting, shooting & fishing? *La chasse* is a common and popular tradition in rural France and as well as providing a hobby for many, they also provide a service in keeping down the numbers of deer and wild boar, which cause many road accidents annually. In the last few years more vegetarian and vegan restaurants have opened away from cities, but tend to be found in tourist areas, and perhaps only during the tourist season. For example, in the 30km radius of our village there are two dedicated vegetarian restaurants, although most restaurants will have vegetarian options on their menu.

It is worth getting a large road map and marking the important facilities you really need, and then narrowing down the areas to say three or four choices. Visiting these areas, staying for three or four days at a time will quickly highlight their appeal or otherwise. You can then either discount the area or decide to focus your search on particular villages or towns. Early spring or late autumn are good times to visit as there are fewer tourists around, it will be cheaper to stay and to travel and you will get a more realistic idea of what the area would be like to live in as a resident . Many restaurants, tourist attractions and shops close after the end of October in some rural tourist areas.

One great bit of advice I can give you from personal experience is to pace yourself. There is nothing more tiring or disheartening than spending all day in a car driving round the countryside to view properties far apart only to find that the area doesn't suit. Before you arrive in France, do as much homework as you can on the internet. Sit down together and make a paper check list as an *aide memoire* of your 'musts' – for example, must be within an hour of an airport, must have access to a big hospital, must have skiing within an hour's drive – whatever your 'must haves'. Limit this to geographical, transport and climate features at this stage – property details will only cloud the issue. We pinpointed the centre of our chosen area then drew a circle on the map narrowing our search area to a radius of 30 kilometres.

For us, we chose Dordogne. It was far enough south for us to be considerably drier and warmer than what we were used to in the UK, although it still gets cold (but short) winters. There are three airports within a 2 hour drive – important for us to get back in a hurry as we had elderly parents in the UK, and children at university; and a good road infrastructure.

Within this department we visited many areas, flying into various airports. Within three years and six visits we had decided on the south of the department. We liked the landscape (which reminded us of Scotland), the towns and the varied nature of farming. It took a further two years and another two visits to narrow down the area to a radius of around 20km. Even then we visited around 30 properties. Some were too large, too small, too much land, not enough land. The internal layout of some didn't suit us, some required complete renovation (optimistically referred to by the *immobilières* (estate agents) as 'requiring decoration' or 'with charm'.

By this time we had made friends with some of the *immobilières*, who knew exactly what we were looking for and who suggested towns that we had not even considered. They would email us immediately any suitable property came on the market and we in turn kept them up to speed with our searching and our own property sale in the UK. Magazines such as French Property News and Living France can be subscribed to and will be filled with beautiful photographs of pretty houses in the sun. With the rise of social media and drone videos many property adverts now feature short videos, which can be very persuasive. Do not let your heart rule your head and remember; no one ever marketed a house with a rainy photograph, but of course it *does* rain in France.

So, from anywhere in the world you can of course search the internet to find *immobilières* or estate agents to search for possible properties in the area you fancy. This will give you an idea of house style and prices and what you can expect for your money. In addition to the usual sites, search the *Safer* site to see not only agricultural or rural properties and businesses for sale, but it also gives you some information on forestry, agriculture, vineyards, equine businesses etc. **www. safer.fr** Another useful site is **www.proprietes-rurales. com** which lists country houses and farms for sale. A more general French site (which also has many properties for sale) is **www.frenchentree.com**

PROPERTY SPECIFICS

We had two lucky escapes before buying our current property. One was advertised with permission for a swimming pool, which just days before the signing of the *compromis de vente* (initial contract) was found *not* to have and no likelihood of ever getting permission. The second, on having the *Dossier de Diagnostic Technique* (DDT) (diagnostic reports) done was found to have termites in a couple of the outbuildings adjoining the house. These checks are very important. In France, having a survey done, as you would in the UK is not a requirement, but the government insists on having the DDT provided by (and paid for by) the vendor and these must be available in full for examination before completion on the property. They cover gas and electrical installation, asbestos, lead, Radon, water quality testing if water is from a spring, termites, septic tanks, energy efficiency and natural and industrial risks and have a six months tenure. If you are looking at properties with swimming pools on them, they also cover any safety features.

Having looked back on our near misses I am glad we didn't proceed with either of them. The first house was too isolated, we would have had to complete the inside of the house and do a lot of building work; we would never have had a pool, which meant our idea of having a *gite* would have been a non starter. The second house was too far from most of the tourist sites, there was far too much land for us to manage, and we later discovered that the local huge pig farm was particularly stinky in late spring and late summer and would not have been pleasant for us or any future guests. (The owner had assured us that the pigs were shortly moving to another location permanently. This was untrue and often you need to ask at the mairie or at the bar to find out the true situation of what Issues a property may have) At the times of viewing however, both houses *seemed* to fit the bill generally and we were very keen to move as soon as possible.

You can search the internet for *immobilières* in your chosen area, but also ask at the local *notaire's* office as often they deal with private sales of property. Agency-specific internet sites such as **www.green-acres.fr** are also useful. Social media is fast becoming an information site and many houses will be placed on local French village sites or property sales groups.

Many sellers advertise houses or sale privately because it is vastly cheaper than using an agent. If you are considering buying privately then remember that you will still have some protection as you have to have a *notaire* carry out the transaction, but remember that there will be no after sales care or service. You will have to do all the negotiating, question asking and verifying of facts, and probably speak pretty good French to ensure there are no misunderstandings regarding boundaries, guarantees, and other important details. The DDT diagnostic reports are still a legal requirement and you need to ensure these are completed in plenty of time and forwarded to the *notaire*.

Use the internet to do a bit of snooping! Google maps and street view are invaluable for seeing what is around a property. And of course you will pass many '*a vendre*' (for sale) signs as you are driving along. You will frequently see the same houses for sale on multiple sites, and for different prices, as French property tends to move more slowly than that in the UK and vendors maximise the chance of selling by using multiple agents. The different prices are also indicative of the different commission rates that the *immobilières* charge (between 5 and 8% of the agreed sale price). And when looking at property schedules or details on the internet, see if the price has the letters FAI or CC beside the price. This means the price includes the agent's fee or commission, but NOT the *notaire*'s fee. Prices with the words *net vendeur* indicate that the buyer is expected to pay the agents commission.

Remember all this when deciding if you really want to purchase a particular property. Are you positive it's the right property? If so, how easily and quickly will it sell in the future? You might be happy to have a property four miles up an isolated unmade road; will others? Will the property still be suitable for you in ten years time? You will have aged, possibly any children will have grown up and moved out. If the property has stairs, will you still manage? An attached annexe that you intend to let out as a *gite* could in future become a better option for you to downsize to, whilst letting the larger property as a holiday let. Think of all possibilities. A property with options is better and cheaper than selling and moving on.

When you visit the *immobilière's* office look around to check they are registered. There should be signs displayed inside the shop with the letters FNAIM, SNPI or UNPI. When arranging to visit property via an *immobilier*, they may ask you to sign a document (*bon de visite*) prior to the visit, to ensure that if you indeed go ahead with purchasing this particular property they will get their commission by being able to prove to the vendor and the *notaire* that they introduced you to the property on a certain date. This is another reason why on the schedule or online advert the location of the house is usually pretty vague, giving only the nearest big town or village. This is to prevent prospective purchasers approaching the vendor direct and thus circumventing the *immobilier*'s fee.

Remember – take the description of the house with a pinch of salt. Estate agents everywhere exaggerate: if you do not want to undertake major construction work on a property, make sure they clearly understand this up front. Their idea of 'habitable' and yours could be very different. If there are English speaking people locally already living in the area, ask for advice. You may learn some nuggets of information that are crucial. For example, *"Oh yes, I know the house. It's the one that they rescued the old lady from last winter when the river flooded."*

Make notes as you view a property for discussion later when you are alone, keep your feet on the ground and a sensible head on. Whilst a property can be renovated, refitted or redecorated, it's much harder to move buildings or change the essential nature of a property. A digital camera or mobile phone are essential tools to take photos both inside the property and of the land and the situation. When you view more than a few properties it can be easy to confuse which one had the basement, what direction South was and how near the house actually is to the road.

Frequently prospective purchasers intend to run *gites* or *chambre d'hôtes* from their properties. The caveat here is to do your homework. A decade ago there was a good market for this with regular business throughout the months of July to September, but in some areas the market has become saturated and hospitality providers are struggling with the competition which in turn drives down prices. Very few people can make enough from a hospitality business to give them a living wage all year, and so it is more sensible to look at any income thus earned as a summer bonus. Additionally; running a *chambre d'hôte* is hard physical work. Are you really thinking of running a hospitality business and keeping animals at the same time? Are you physically fit enough and can you maintain your smile and pleasant nature even at the end of a very long day with difficult guests? *Gites* and *Chambres d'Hotes* are very popular with the French, so consider advertising widely and in French.

Many rural French properties come with a larger parcel of land than we are used to in the crowded UK. It is very easy to be seduced by this. Whilst the dream of becoming lord of the manor and surveying your hectares from the top of a hill is instantly appealing, land needs work and maintenance. If you are used to thinking in acres then a rough guide is 2.5 acres to a hectare. A couple of acres is enough to become fairly self sufficient, perhaps with an additional acre or two of woodlands for fuel, but unless you are a wildlife enthusiast who is happy for woodland or pasture to have minimal management, any more than this is for real farm enthusiasts, and beyond the scope of this book.

Land is relatively cheap in France and whilst this is a great temptation, consider realistically about how you are going to utilise it. Even a couple of acres of lawn require mowing at least once a week and for many more weeks than in the UK. Here in Dordogne my first cut is in February and the last cut is a week before Christmas! I have invested in a good ride-on mulching mower, which is a necessary expense to reduce both time and effort. If you have fields to cut, then perhaps a tractor with a pull behind topper will be more use than a ride-on designed for flat lawns. Hedges need an annual cut which takes me and my husband a whole week to do properly. Also consider that, in the future if you retire to something smaller or for whatever reason have to sell, a property with a lot of land is going to be far harder to sell. On a large acreage you may also have to acquire approval from *SAFER*, who may offer the land locally for sale to existing local farmers. Normally, this is just a formality.

So, let's look at the land on offer. If the house is surrounded by or attached to the land this historically has been poorer quality land. Most farm buildings (*les dépendences*) have been built for easy access to roads or tracks and many are sited on thin or rocky land with good drainage. Is there any natural water; whether this is a stream, spring or pond? Bear in mind any flooding risks and whether the supply is constant all year. Asking at the *mairie* may provide a more truthful answer than asking the vendor. Is the area susceptible to drought restrictions? In early 2019, over 11 departments out of a total of 96 mainland departments were already on drought restrictions before June. Drought affects not only crop production, but also pasture and hay crops meaning costs for animal feeding will go up. Drought also damages soil structure.

Some rural houses or fermettes have water tanks buried in the ground which collect rainwater from the roofs. These tank systems are expensive to put in, so are a bonus if in good condition. They can be adapted to use rainwater for flushing toilets as well as for cleaning the car and watering potagers or irrigation of larger areas of land. A roof can collect a fair bit of water – one mm of rainfall can give you one litre per square metre of roof surface. Remember and clearly mark the pipes coming from any rainwater harvesting system – this is now a legal requirement.

The table below gives you a rough idea of what a certain amount of land will allow you to do.

0.5 acre	Vegetable garden, small orchard, chickens, bees.
2 acres	As 0.5 above, & 2 seasonal lambs or 2 young pigs, ducks, polytunnel.
5 acres (2 hectares)	As 0.5 above, & woodland for fuel, polytunnel, small starter flock of breeding sheep or goats
10 acres (4 hectares)	As 5 acres above & larger woodlands, hay meadow, walnuts, small vineyard, 2 breeding sows.

Is the land adequately fenced? Normally fields for arable cultivation are unfenced. However, if you intend to grow vegetables, have an orchard or keep livestock, it is the landowner who is responsible for keeping trespassing animals OUT of his property, and keeping your own animals in. A quick walk through the land if you are seriously interested is an opportunity to discover any standing water, dumped rubbish, abandoned cars etc. Remember if you are going to grow fruit or vegetables to check for possible frost pockets in gullies and hollows. Steep hillsides will pose a problem for machinery use and possible erosion or localised flooding.

A general inspection of surrounding farmland will indicate what the land is generally used for. If there are a variety of crops then this indicates good soil. Grazing for dairy or beef cattle or fattening lambs indicates a slightly poorer soil. If the land is scrubby then ask yourself as a smallholder how you will utilise it. Having a good look at land is very informative, for example the meanders of the rivers Dordogne or Vezere. On one side is often a tall limestone cliff, whilst on the other side a flat floodplain. On the steep side, the land is often inaccessible and hard and thankless to cultivate and dangerous for animals to be farmed. The other side of the river bordered by floodplains is usually fertile and productive with crops such as vegetables, tobacco, cereals and sunflowers, but it is still a floodplain and liable to flooding. Farms and older houses are set back on higher ground for a reason.

Hillsides in hot areas are usually the domain of wine growers but this is a specialised field. Unless you intend to rent your land to a neighbouring vineyard are you really going to embark on a project that could fail and swallow a lot of your hard earned capital?

Checking the quality of the land is important. The best soil is a medium loamy friable soil. An ideal pH level for growing most crops is around 5.5 – 7.5. This can be ascertained by a small sample in cheap-to-buy test tube kits. You add a small amount of water, shake and read the colour change on the chart. Using your eyes, you may see lying water on the soil surface or perhaps the presence of rushes, which indicate water logging and/or acid soil. A low pH (from 7.0 downwards to 4.0) indicates acid soil, a high pH (from 7.5 upwards to around 9.0) alkalinity. Remember pH can vary from one end of a field to another. Applying lime can 'sweeten' acidic soil and the addition of humus, manure and leaf mould can add body to thin, dry soils. In Cornwall we had shallow, shillety soil in one area (rubbish for apples but great for cherries) and deep loam in another area (perfect for apples and vegetables).

The view from a high hilltop is breathtaking, but the exposure to winds will not do much good to fruit trees, poly-tunnels and may bring some damage to buildings and woodlands. Conversely, properties in valleys could have issues with high water levels and damp, and lack of sunlight and air movement in valley floors may lead to problems with mildew and crops or fruit failing to ripen.

Examine outbuildings to assess how much work/money is required to bring them up to scratch, whether they have an existing electrical or water supply and look at what your intended use could be. Disused buildings could be transformed into holiday accommodation if they are sound enough, you have a vast pot of cash and your *mairie* is willing. You will need to complete a planning application or any major works to buildings or for the installation of an In-ground swimming pool (*déclaration préalable de travaux*).

Is the neighbouring land also agricultural or has it been upgraded to constructible land (thus increasing its monetary value). If so, bear in mind that you may lose your view to a newly built house or houses and the associated noise and nuisance that accompany this. A quick enquiry at the local *mairie* should put your mind at rest, but this should be confirmed during any subsequent searches if you decide to go ahead with a purchase. Any woodlands either on the property or surrounding will possibly be home to wildlife such as wild boar, deer, foxes and badgers, which may also mean visitation by the *chasse* with dogs and guns in season or damage to important crops you may be planning.

Check the electrical supply. When (if ever) has the supply been checked or rewired? Having a look at the fuse box is a good indicator of how things stand. Is the property on three phase or single phase or a combination of both? If the property has been a farmhouse or farm building it may be worth having a professional electrician round to give you his thoughts. Three phase (*triphasé*) simply means you have three live terminals and a bigger fuse box. You also pay a higher standing charge for three phase as the supply enters at 15kVA.

Single phase supply can be installed by a qualified electrician as this is the most suitable for domestic use. You will need to select the most appropriate kVA supply for the size of your dwelling by working out the maximum capacity load you would use. The most common units are 6kVA, 9kVA & 12kVA. EDF will help you and have an English speaking helpline. The lower the kVA the lower the standing charge, BUT don't go too low or you risk tripping out on a daily basis. It is usually free to have this increased.

Is there a well or borehole (*forage*)? Whilst 99% of properties have a mains water supply, some rely on boreholes or wells. If this is your only source of water and you wish to use this as drinking water you must ensure that it complies with current regulations as to public health. Technically, you should also inform your *mairie* that your supply is fit for consumption. Ask the vendor to show you any water testing results and their bills. All water, whether from mains or wells or boreholes '*should*' be metered and paid for (yes, even if it is from your own property). If it appears that the supply is not, then taking a sample for analysis at a large pharmacy or to your *Laboratoire Municipal et Regional d'Analyses* should determine whether it is safe to drink. If the water is deemed unsafe for human consumption, then I personally would not use this for animal use; especially if I intended butchering those animals for food.

Finally, try and find out the tax liability for the property you are looking at. A big house with land and a pool will have a considerable tax bill at the end of the year. There are two taxes to consider. *Taxe d'habitation* and *Taxe foncières*. The first is a French residency tax and is liable even if you are not normally resident in the property. The second is a French property ownership tax, which finances local services in the commune and department in which the property is located and is currently being phased out gradually. If you make any major improvements to the property you must inform the tax authorities (or your jealous neighbour will), who will then arrange for the property to be revalued.

THE BUYING PROCESS

Let's imagine you find a suitable property. You have gone back for a longer, more thorough look, possibly taking someone with you who speaks fluent French to clarify details of boundaries, fixtures and fittings, furniture and appliances that will be left. You've ascertained the time frame that the vendor will leave and that you could take possession, you've discounted any other properties, checked your finances are in place and are ready to make an offer. Have you discussed with an expert in house purchase and inheritance law the various methods to purchase this property? If you are a non-married couple, same-sex couple or have children from a previous relationship you are advised to take advice before proceeding.

Negotiating to get a better deal is worth trying. The *notaire's* fees are not negotiable, but if the property has been on the market for more than six months and you've checked online for similar properties in the same area you can get a feel for what the property is worth. You could explain to the agent that although you really like the property, and of course being a cash buyer you can move immediately, you feel that the property is overpriced or the land of poorer quality than the price suggests, or the fosse requires updating to conform or whatever your particular bargaining chip may be. Taking all this into account you'd like to offer XXXX euros. A starting bid of around 20% less could be considered cheeky, (but normal for French buyers), but you don't know the vendor's situation. If the sale is an estate sale, then the family may be in uproar and demand more money or be keen to accept a reasonable offer to be rid of the burden and be able to split the inheritance. This could go on for years. The game of bluff must be carried out with caution; lest you offend the vendor and he refuses to entertain you at all.

Ensure any offer includes any conditions (*clauses suspensives*), such as the pre-approval of mortgage on the property or the completion of a sale of property elsewhere; any remedial works to be completed and paid for by the vendor, and details of what fixtures and fittings you wish included (eg woodburners, gite furniture, tractors).

You can either make the offer verbally to the vendor, or if you are dealing with an *immobilier*, with their assistance, and preferably by email or in writing to ensure everyone knows exactly what is being offered and for what.

Normally the vendors will have chosen a local *notaire* for their side of the transaction and there is nothing to stop you using the same *notaire*. The notaire does not favour either party of the transaction, he acts for the government to ensure the property is transacted legally and that the correct taxes are paid to the government. His fee is set by the government and he will not normally give you advice. Many speak very good English, but if you prefer a different *notaire*, this is your prerogative. The advantages of keeping the same *notaire* mean that all the documents are kept in the same place and there are fewer delays.

If you are concerned at all about understanding the procedure or certain technical points that come up during the sale you can always employ a French Solicitor to examine the documents. If this professional is also English or speaks very good English this can only help you fully understand the details of your purchase and alert you to any causes for concern. He can also suggest the inclusion of any clauses or conditions he thinks necessary (*clauses suspensives*).

So, as to buying process; once you have found a property you like and agree a price and a tentative date of entry with the seller, you have assured yourself as to what is included in the purchase and what the boundaries and land are you or the agent you are buying through contacts the *notaire* who draws you up a contract; a *compromis de vente* (an initial agreement of sale).

You do not have to be present at the *notaire's* to sign this if you are still resident in the UK, it can be done by email and post. If the notaire posts you the *compromis,* (by recorded delivery) the 10 day cooling off period begins the day following his signed receipt of the contract.

At the end of the ten days the purchaser and vendor have a legal agreement and both have to forward 10% of the price to the *notaire's* bank account. This deposit will become forfeit if either party then pull out of the sale (The percentage is often negotiable but, under most circumstances, a deposit of an agreed amount or percentage is expected).

It takes an average of three to four months to complete the sign, during which all the searches and diagnostics are carried out. Try to avoid the month of August as the French system grinds to a halt as everyone goes *en vacances.* The balance of the outstanding property price plus the *notaire's* fee then have to be deposited in the *notaire's* bank account a few days before the completion date. All *notaires* will ensure you have an assurance policy in place for the new property before you actually take possession. Note: the French system refers to assurance, not insurance. You must have a policy in place to cover buildings insurance, and cover for natural disasters such as earthquakes, forest fires and floods. This will be required whether your property is your main home, a holiday home or a buy to let.

The *notaire* prepares the *acte de vente* and both parties are invited to sign at the *notaire's* office. You can instruct a Power of Attorney to sign on your behalf if you cannot attend in person. Ensure you check the property shortly before signing the *acte* to ensure it is in satisfactory condition and that any repairs have been completed. Normally, this would be done the day before the *acte* is signed.

The *notaire* will read through the *acte de vente* (In French) and both vendor and purchaser are asked to initial each page. The last page is signed, together with the words "*bon pour accord*" written besides the signature, to show that you understand and accept the terms of the document. You may be able to ask for an advance (electronic) copy of the *acte* for translation into English. This is fast to do with modern translation software.

The keys are handed over and that's it – you're the proud new owner of your very own house in France. So, you've spent a fortune and are probably exhausted, what better way to celebrate your new life than with a simple, typically French meal.

MONEY, BANKS & TAX

You will need to ensure you have a bank account in France before you arrange to pay your deposit to the *notaire* as part of any house purchase. There are many banks who can be approached and with a letter from the *notaire* to give proof of the intended purchase and a sum of money to be deposited you can start the process whilst on a visit.

Choose a bank that can deal with all your requirements. If there is a branch near to your intended new home, so much the better; especially if any of the staff speak English. Again, ask around for recommendations. Personal banking is not free in France, so ensure you know what various charges are for.

Remember also, that transfers from a UK bank to a French bank may take longer than is customary in the UK. Never write a cheque on a French bank if you do not have the funds to cover it. This is a serious offence in France and could result in you being blacklisted by the Bank of France. Also, remember that once a cheque is given, it does not have to be banked immediately, and it is valid for 12 months and eight days. It is also very hard to cancel a cheque.

At some point during your move to France you will have to exchange currency from either sterling or another currency (unless you are already using the euro within Europe). Exchange rates fluctuate constantly, particularly so in times of economic uncertainty and political upheaval. Consequently these fluctuations can make a significant difference between what you expect to receive for your transfer and what you actually end up with. I know this from personal experience.

If sterling is strong against the euro your pounds will buy more euros and conversely if sterling is low/weak they'll buy less. Rate movements over just a few days, hours or even minutes can make a massive difference when you are exchanging large sums of money.

However, it is possible to mitigate some of the risk by using the services of a regulated currency broker rather than transferring directly from your bank. For example, a currency broker will allow you to 'lock in' an exchange rate for up to 12 months (a deposit will be payable up front to secure this rate). This means for instance you can fix your rate for your balance transfer after you have signed your *Compromis de Vente* - thus guaranteeing your purchase price in Sterling regardless of what goes on in the wider world during the months that you are waiting to complete your purchase.

Another very popular option is a product called a Limit Order. Quite simply you tell your broker your target rate (i.e. the rate at which you will be happy to buy euros) and how much you need to buy – and when your target rate is reached, your euros are automatically purchased. The beauty of a

Limit Order is that you can change or amend at any point up until the point that the rate triggers the purchase – so if you are running out of time or you feel you have set the target rate too high you can change this.

If you need to act swiftly then you need a Spot Trade – this is a buy now pay now option to secure you the best rate at that moment. Using a foreign exchange specialist instead of your bank will also get you a preferential exchange rate. There are many firms offering money transfer services so you need to research what is best for you. For my house purchase I used Currencies Direct and found their service and advice invaluable: **www.currenciesdirect.com** and for small regular transfers I use **www.transferwise.com.**

Naturally, the two things in life which you cannot avoid are death and taxes. Death is considered at the end of this book in the appendices, but we will touch on tax here.

Whatever your situation when moving permanently to France you need to consider that you will have to be accountable and have a tax return of some sort; whether you are retired, inactive or working. As the tax system is complicated and every individual has unique circumstances, I would start to gather information and advice at least six months before purchasing a property in France.

Once you live in France permanently for 183 days or more (whether you are working or not); carry on a professional activity (whether self employed or employed) and have a centre of economic interest in France, you are considered to be fiscally resident – ie liable to register for and pay tax. The tax year runs from January 1st to December 31st. Ensure you have fully explored and understand the implications of the tax situation as it applies to you and its impact on your tax situation in the UK, before you make any property purchase.

Capital Gains Tax is liable on the profit made from the sale of property; particularly (but not exclusively) from the sale of second homes. If you are resident in France you are exempt from paying CGT (*impôt sur les plus-values*) from the sale of your main house. However, you are liable for CGT if you sell a second home unless you can prove ownership of that home for 30 years or more.

If selling a second home, CGT is calculated by deducting the purchase price from the sale price. The costs incurred during the sale and any renovations or improvements can also be deducted (remember and save all those official receipts and invoices and ensure they are correct and have Siren numbers). The *notaire* will calculate the correct CGT due and deduct this before passing the proceeds of the sale to the vendor.

In the event that you do not sell your first (main) home before moving to your new home, you may be allowed an 18 month period of grace to avoid CGT.

PAPERWORK

The golden rule is bring everything you can. For yourself, you will need birth, marriage and divorce certificates. This is especially important for women, as in France women retain their maiden name; so if you have married and perhaps subsequently divorced and even re-married you must be able to produce paperwork to prove the timeline and paper trail. If you hold a British passport ensure you have at least a full 12 months validity left on it.

For future car insurance bring all your old car insurance records for at least 13 years. Yes, in France to get the best deals you need to be able to prove you have had car insurance (not no-claims bonus) for a run of 13 years without a break.

If you intend to work in France for an employer then, all your certificates in full, and any proof of courses that you have attended if they affect your ability to do your job together with training schedules, apprenticeships etc. are all the sort of thing that you need to bring with you.

You will also have to have a checklist of places you need to change address etc. Whilst this may seem straightforward, remember you will also have to inform the DWP, tax office, your GP and dentist, pension provider, and if you wish to continue to vote in France, your local authority. A wise precaution is to check how long your UK full driving license is valid for and if you need to renew it, do so many months before your anticipated move. Do not inform the DVLA of your new address in France until you are ready to apply for a new French driving license. As I write this book, the French online system (new in 2019) is overloaded and it may be some time before new applications are processed. The French government are allowing UK license holders to use UK licenses whilst the backlog and application process are streamlined.

What is your situation regarding wills and inheritance? If you don't know then you had better start thinking about it prior to moving to France, especially if you do not have a straightforward situation. You must mention to your lawyer or *notaire* if you have a will in place in another country (ie the UK). France recognises valid UK made wills which covers all assets (including French ones). A new French will CANCEL out an existing UK will and vice versa, so you need to make sure you are clear about what you wish to happen to your assets after your death, especially if you have remarried or have children by a previous marriage. In France you cannot disinherit your children, so if you wish to exclude a child you must consult a solicitor and have a UK will written.

If you wish to maintain a will made in your previous country (e.g., the UK) you must specifically mention this in the will, by stating, for example, "I cancel every previous will except my will made on (insert date) in (insert country – either UK or France) that I specifically maintain".

A single will can cover assets held in both countries, however, this can lead to difficulties as the inheritance and tax laws in the two countries will be different. French law does not recognise the powers of an executor of a will, nor does it accept assets held in trust. Even if the beneficiaries of assets in France are children, a discretionary trust is heavily taxed.

French law recognises four types of will; an *authentique* will which is dictated by the person making the will to a French *notaire*, who writes it up and is then signed in the presence of two witnesses. A *mystique* will is written in private by the person making the will, sealed in an envelope and handed to the notaire in the presence of two witnesses. An international will can be typed or hand written by the person making the will and signed before two witnesses and the *notaire*; and finally an *olographe* will (the most common in France) is written entirely by hand by the testator (person making the will) and is dated and signed by the testator. It does not require any witnesses and can be written in any language. The original will should be photocopied, the original being given to the *notaire* to be kept and registered at the French national registry of wills (*fichier central des dispositions de dernieres volontés*). This is not a legal condition, but to ensure that a will can be found. Copies of the will can be kept by the testator or distributed to those future beneficiaries.

A quick word regarding the French Postal System. Mail is delivered six days a week. You can of course visit a branch of *La Poste* to post any mail or parcels you have, but be aware that important or valuable letters are best sent *Recommandée* rather than simply by first or second class mail.

You can also buy stamps at tabacs (newsagents). There are three standard tariffs – *Prioritaire* (first class), *Lettre Verte* (2nd class) and *Ecopoli* (a cheap bulk mail service that can be a bit slow). *Lettre Suivi* is a letter sent with tracking. For international mail here it is standard International postage. There is also a recorded delivery service both nationally and internationally. Recorded delivery is widespread in France where it is recommended for legal and valuable documents, contracts and correspondence with official bodies. You have the choice of *Recommandée avec accuse de reception* (with an acknowledgement of receipt signed by the recipient) or *Recommandée* (recipient signs for the letter, but the sender gets no acknowledgement. You can buy three levels of cover for either method of recorded delivery.

REMOVALS

There are many specialised firms that regularly travel between the UK and France. If you are relocating from a different European country or from further afield, it pays to make enquiries some months in advance. You will have to assess as carefully as possible the size of your load (in cubic metres) and whether you need a specialist for certain items (eg vehicles, antiques or large machinery).

Also, take into account relocation at summer peak periods – this may attract a premium on the cost; and whether you will accept a contract as a part-load or full load. Most companies give you the option to pack yourself, although ask what the implications on insurance this may have. You also need to decide between 'new for old' insurance cover or indemnity cover (replace like for like, including age etc). Ensure you understand what Goods in Transit insurance is and what level of cover is sufficient for your needs. Read through any documentation before agreeing and clarify any points which are unclear.

If you are packing yourself, ensure you don't skimp on the packaging. Having a colour code and number system for boxes is a particular tip I pass to you. I've owned 12 houses so far, and moving house is a streamlined operation for me as I've done it so often. I mark boxes with coloured thick markers – a different colour for each room, and as I pack and seal a box it's giving a running number. So a box marked in red with no 2 means this is to go to the kitchen and is no 2 in the sequence for that room. I also keep a checklist of rooms and what is in the numbered boxes (eg books, shoes, ornaments etc). When I get to the new destination it's simple for me to open the more immediate boxes to find the kettle, cups, cutlery etc, and leave the less pressing unboxing for the future. Of course if you have a partner, this system only works if they follow it too.

Ensure your chosen company have your full new address, mobile phone number and anything they need to know on arrival, such as height restrictions, narrow access, and whether there are stairs at the new property. If you need storage at either the start of your removal or at the end, again, make your enquiries many months in advance. Good removal firms are booked up quickly and you must try to organise your move at least four months in advance. Good communication with the firm is essential. Moving house is traumatic enough without the frustration of poor communication.

A sensible idea is to pack a case with overnight things, a change of clothes, any medication you require and keep this with you. I'd also keep any official documentation, birth certificates, insurances and keys in a case with you. Ensure you have a charger and a spare mobile phone battery with you, and sufficient change in euros for any toll roads.

Whilst the United Kingdom was a member of the European Union it was permitted to bring certain plants into France.

There are a few exceptions to this and you can find out more by visiting the Defra website. At this time it is unclear what rules will be established regarding plant importation.

If you are transporting animals as part of your relocation, please ensure you begin this process some four months in advance of the move. Start by asking your Veterinary Surgeon what the latest requirements regarding microchips, vaccinations and paperwork. Also, think about how you are going to transport animals. It's a different kettle of fish relocating pedigree sheep than it is for a cat or dog. (See the section on animals below.)

HEALTH INSURANCE

The healthcare system in France is complex and expensive, but it is a wonder and offers fantastic service that beats the British NHS hands down in my opinion. There are few waiting lists and the services generally offered are high quality. The French are thought to be a little obsessive regarding health, but as such many screening services to prevent illness are free and regular, such as cancer screening and well person clinics.

The system is known as '*l'assurance maladie*' and to be able to take advantage of it you need to be resident for at least three months and to live in France or at least six months a year. During the Transition Period until December 2020 retired Britons living in France can apply and if already resident here can use the S1 form if they are in receipt of a UK state pension. You must contact the DWP for details of this. It is unclear at the time this book went to press what the arrangements will be after the Transition Period, but it can be assumed that if you are NOT covered by the Withdrawal Agreement that you will have to take advice on how to obtain health cover in France.

If you are taking up employment in France or becoming self employed there are different organisations which will sort your healthcare depending on your type of business and legal set up. Normally, we understand, unless you are working or have a business, it takes 5 years before you can apply for a *carte Vitale* and you must be able to prove that you have had full private health insurance in the interim.

Eventually, you will receive your *carte Vitale*, which entitles you to healthcare, and you may need top up insurance in the form of a *mutuelle*. You need to present both any time you visit a healthcare professional, and at the pharmacy too. Pre existing conditions and age are NOT penalised here in France. Think of your *carte Vitale* as your National Insurance card. It ensures you are in the social security system. It works as a reimbursement system. You visit the doctor and pay up front; then your *carte Vitale* is swiped and the government reimburses some or all of the cost to your bank account. How much you get back is dependent on the treatment or action. The rest of the cost can be paid by your *mutuelle* (top-up health assurance).

You cannot use a European Health Insurance Card (EHIC car; formerly known as an E111 card) for medical treatment if you are permanently living in France and so you need to prepare in advance to provide for your care.

You can find out more about eligibility and obtaining French healthcare by visiting the Ameli website:

www.ameli.fr/assure/english-pages

CONNECTING SERVICES

Naturally, you will want your water, electricity, gas (if you have it) and phone connected as soon as possible. Before the big move liaise either with the seller or the agent to start the process of switching over. If your commune deals with your water supply, a trip to the *mairie* after you sign for your house with a water meter reading may be necessary. Take a reading of your electricity and contact the supplier. If this is EDF (*Électricité de France*, they have an English speaking helpline). Getting a phone line installed or connected can be lengthy to arrange, so you may need to have a mobile phone until this can be connected or transferred. Ensure you read or have someone to translate the terms and conditions of the contract very carefully. I got stung from a firm for two years and was paying for services I didn't need, that I was not told about when arranging the contract.

ANIMALS

If you intend to bring family pets from the UK to France you need to think about two things – are you going to take them back to the UK at any point (for holidays or relocation), and the considerable cost and paperwork involved in bringing an animal over in the first place.

At present all pets (dogs, cats and ferrets) need a Pet Passport to both leave and then re-enter the UK. All pets must be in good health, have up to date vaccinations, have microchips implanted and have a rabies vaccination.

If you are bringing your pet in your vehicle via the ferry or Eurotunnel, then consider how the pet is going to tolerate the strange routine and perhaps speak to your vet about tranquilisers. Pets travel better on an empty stomach and need plenty of ventilation. On arrival at your ferry port or train departure point your paperwork will be checked and pets scanned with a handheld scanner. This is a good time to give dogs a final walk pre-boarding. Ferry companies that allow dogs on board, offer visits whilst on the journey to check your pet, and some offer pet friendly cabins, but obviously you will have to check each individual company for information and pricing for this. The Dover-Calais route do not as the journey time is just one - two hours).

If you intend to return to the UK at any time with your dog, you must ensure that his rabies vaccination is up to date and that he has visited a veterinary surgeon between 24-48 hours before travel to be checked for ticks and to have a tapeworm treatment. Your vet will give you paperwork attesting to this procedure, in addition to your Pet Passport for the animal which you need to accompany you when travelling with the dog.

Certain breeds of dog are considered dangerous animals and France has prohibited them being brought into the country. Please check with your vet if you think your dog may be included in the breeds or types prohibited.

Ferrets must also have a Pet Passport, but rodents including rabbits are currently free of restrictions, but you really do need to check with your vet, and also with your transport provider if they will accept such animals and what the conditions of transport are and the fee.

If you intend to relocate any horses, farm animals or poultry you must check the up-to-date regulations for such. At this time quarantine regulations may be enforced for any animal not qualifying for a Pet Passport. Horses, donkeys and mules all require to be microchipped and have a horse passport at the least.

Animals such as poultry, sheep or cattle are NOT pets. There are specific rules and regulations to cover moving these permanently from the UK to France. At the present, pre-Brexit rules state that you will require that all sheep, pigs, goats and cattle require ear tags, many require blood tests (check what you will require for each species as this differs). The APHA section of Defra is a good place to start researching, but if you already have farm animals then your large animals vet should be able to give you advice also. (The latest advice is included in Animal & Plant Health Agency APHA Briefing Note 17/18 issued October 2018).

In the event the UK leaves the EU with no agreement, then you must consult the APHA website or ask your large animal Veterinary Surgeon for the most up-to-date rules. Frankly, it's going to be expensive and a paperwork and logistical nightmare.

Tarte aux mirabelles.

For the pastry: Or cheat and buy ready made pate sucree
113g unsalted soft butter
50g sugar
1 egg
Pinch salt
225g plain flour
2 tbsps water
For the filling:
125g ground almonds
125g soft butter
125g caster sugar
25g plain or mixed plain & SR flour
1 egg
1 cup chopped pitted mirabelles (or peaches or nectarines)

Pastry: In a mixer, whip the soft butter until pale, add the yolks one at a time and beat in well. Add the salt, sugar, flour and mix briefly. Add a little water to make a stiff dough. Dump the dough onto a floured surface and gather into a ball. Wrap with plastic wrap and refrigerate for an hour.

Filling: Place the ground almonds, flour & sugar into a bowl, mix. Beat the egg and add to this mix with the butter. Mix completely. Leave to stand for 5 minutes. Meanwhile, preheat oven to medium heat (150 C/350 F).

Butter the bottom and sides of the pie dish. Roll the pastry onto a floured surface and roll till uniformly around 2mm thick. Line the pie dish with the rolled out pastry. Spread the filling onto the pastry base and push the prepared fruit into the mix, submerging it slightly.

Bake for around 30 minutes until golden.

Chapter 3
Land Usage

In general, there are some common sense facts you need to remember, when looking at land utilisation. Firstly, land that surrounds or is attached to your house is a more attractive option than a separate parcel of land 'down the road'. It will also be much easier to sell if and when you decide to sell your property. Square shaped fields are easier to use and more efficient than long, narrow ones. Steep slopes can be dangerous when using standard tractors or machinery, and can be eroded by wind or heavy rainfall. Close proximity to any large blocks of housing may encourage stray dogs, stray children and dumping of garden waste or other refuse. A source of clean water on site is usually an added bonus (if controlled or managed), and checking that neighbouring properties or farms are tidy or well kept will give you peace of mind. Also, although expensive; fencing is your friend.

GARDENS

I've always loved gardening and I often look back and reminisce about my previous gardens, and this knowledge is a great tool when setting out a garden that is new to you. You *know* not to plant trees too close to the house or that elaborate planting schemes require more time and effort than you really have time for. However, a new and unknown garden can be a daunting prospect for most home owners.

As an ex garden designer my advice for someone taking over a house with an unknown garden is to keep a diary and take photos for a year. What does it look like in winter? Take a photo and make notes beside this in the diary. Useful notes; such as nice open view in winter, but not much colour; or really bright orange rudbeckias in the middle of a mainly purple/pink planted border. Noting any frost pockets, shady spots and areas prone to flooding will all be useful when planning your garden.

As you move through your first year remove anything obviously dead or dying, repair any fencing you wish to keep, look at existing lighting if there is any; water reclamation (everyone should be doing this now) and think about areas that would make nice seating areas, dining areas, a place to sit and enjoy that fabulous view. You get the picture. Try and name any shrubs or perennials or bulbs that appear. This is not a waste of time! I had a non-descript leggy bush in a border that was swamped by euphorbias and a Choisya. I had earmarked the mystery bush for removal and was about to start chopping it this summer to give the others a bit more room when I spotted the most fantastic flower on it! A quick internet trawl revealed it was a Fejoia - a remarkable South American fruiting bush (and fairly expensive to buy). So, it's now staying, and the euphorbia is going. There are plenty of them self seeded all over the flower beds, so it's not a loss. The Choisya has been clipped back to a more pleasing shape and the bed is happily providing colour and scent.

If you are a little afraid of garden design, please don't be. The basics are pretty simple and often guided by the shape of your plot and the soil/climate you have. Now, in France the climate varies, but you will know if you have a sunny dry plot or a shady damp one. That's all you need to know. You may have a terrace house with a back garden, a farmhouse with no real garden, but lots of land, so traditional garden design is out the window, as everyone will have a unique space. Think about where the sun comes, and where you'd like to have a seating/lounging area. Somewhere near the house is nice. A dining area would benefit from being under a lofty shady tree or you could have a pergola of sorts draped with vines which sadly attract hornets (*frelons*) or wisteria if you fancied this. Lighting, whether it is mains or solar will extend the time you can spend in your outdoor rooms and can be functional, such as path markers, or fairytale such as tiny lights hanging from the trees or shrubs.

Your allocated budget is going to be a big factor in deciding what your priorities in the garden are. Hard landscaping can be expensive in France. Different paving breaks up a large area with slabs or flagstones being nicer to walk on with bare feet than gravel (although more expensive), and if set into the grass are easy to mow straight over. I intend to extend a small slabbed patio outside my family gite next spring with a checkerboard of yellow and pinkish slabs. The simple addition of bi-coloured black and white paint tin lids will not only give me an extended patio, but at the same time provide an outdoor draughts or checkers set!

I love scent, so I have jasmine, lavender, herbs, scented roses, and winter flowering Eleagnus. In little pockets I have Lily of the Valley, and along my fences I have Clematis Montana, many of which are scented. Raised flower beds are easier to manage with taller, bushier plants at the back and smaller plants and bulbs to the front. I like to slip in the odd vegetable such as tomato plants, swiss chard and alpine strawberries in amongst the planting as little surprises for any visiting kids.

You need to pick a few stalwarts for every time of the year and try a colour combination such as blues, lilac & pinks with silvery foliage; or yellow, oranges & red with gold foliage. If you're feeling adventurous white as a predominant colour picked out with the occasional accent colour (yellow, or purple or pink) can work well. There is a white garden at Eyrignac in Dordogne that works well with a background of dark green foliage. Borrow some library books or visit some gardens to see what you like and ask other gardeners. All gardeners love to talk about their gardens and may press rooted cuttings or seeds upon you as you leave. Propagating some plants can be really easy and save you a fortune. My children will roll their eyes and tell how every time we visited a National Trust garden when back in the UK I'd have a tiny plastic bag and scissors in my bag.

Cuttings can be one of the easiest ways to propagate hardy perennials. I've learned, for example, that any of the salvias and sages root readily from cuttings just taken and thrust into a shady spot in the ground. Roses will root if 6 inch long cuttings have their bottom leaves removed and thrust into the ground too! I only tried this for the first time two years ago and I've enjoyed a good success rate, especially if you pick cuttings as thick as a pencil.

Lots of other hardy perennials can be split with a spade in spring or autumn and replanted elsewhere in the garden. Euphorbias and fennel (and many other plants) will seed everywhere; just remove the ones in the wrong places.

I have stopped using old ice lolly sticks as name markers, which discolour so much after a season that I can't read what I've written on them. Recycling old yoghurt pots and cutting into strips provides me with a good few eco labels. Making a garden map will give you a more permanent *aide memoire*.

A compost heap or two can be hidden away not too far from your working garden, and water reclamation can be attractive to collect water from the house roof or practical, in the form of 1000L IBC containers at the rear of the garage or outbuildings.

One final word about gardening. From August onwards the soil may become host to minute little mites which can get on your hands and clothes and cause intense itching and irritation which may last a few days. These *aoûtats* can infest pet paws too! Ask at your pharmacy for products to sooth the itch and to prevent biting.

WOODLANDS

Having a woodland or small forest is a dream for many people moving to the countryside. The amenity use of such an area of land is hugely beneficial, but woodland, like any other land requires management. In addition to providing fuel for winter burning, they also provide a habitat for flora and fauna which could provide a useful amount of game for the freezer, a home for many creatures and plants and a welcome and beautiful environment for refreshment of the body and mind. Trees also provide a natural windbreak and an attractive backdrop for any home.

The variety of species calling your woodlands home will naturally depend on whereabouts in this country you reside. But you can almost certainly include oak and sweet chestnut amongst the trees, red squirrels, rabbits and deer amongst the mammals and a variety of bats, birds and insects too. In some heavily wooded areas wild boar are common, and certainly in the more remote areas wolves are beginning to be seen again, although these are rare.

To supply your woodburner or fire a hectare of land can supply enough wood fuel to last two years, but once cut you will have to wait another five years minimum to cut for wood fuel again. If you have more woodlands than this, a system of coppicing or replanting can provide a sustainable source indefinitely. Young saplings can be interplanted amongst older trees to continue the cycle and new species can be introduced to vary a monoculture and provide a more dynamic micro environment. Softwoods such as hybrid poplar are fast growers but will still take up to six years to grow large enough for harvesting for firewood, and much cut timber will take two years or more to season before you can use them in the woodburner. Specialist tree nurseries (*pépinières*) sell bundles of young whips for you to add to your woodlands, but remember that a young tree is a delicious food source for browsers such as deer, rabbits and hares. Whips can be planted in spring or autumn with tree guards to protect them.

The best species to plant for wood fuel tend to be oak, ash, sweet chestnut, beech and acacia, but you will have to check what grows in your particular region and order accordingly. Stands of fast growing poplar are a common sight whilst driving along roads, but these do not produce the best firewood.

I added a few elder and hazel bushes in my woodlands as I enjoy using the elderflower for cordial and wine and enjoy adding hazel nuts to my crop of walnuts just to vary the nuts I enjoy. I also added some holly for festive reasons and existing sloes will provide for that old stalwart, sloe gin or vodka (see the recipe at the end of this chapter). Pine trees are softwood and need stacking also for at least two years to avoid the resin sticking inside your chimney and burn very quickly.

Remember that cutting and stacking trees for firewood is a dangerous and labour intensive job. If you intend to tackle this yourself, you need to invest in a good chainsaw (*tronçonneuse*), and good personal protective equipment, (PPE) such as helmet with visor, Kevlar trousers or dungarees, stout gloves and boots. The best time for this job is between autumn and late spring and the wood must be left to season for at least two years. Never harvest firewood alone in case of accidents and stop when you are tired.

Coppicing involves cutting certain tree varieties down to the stool from which they grow. This way you can harvest timber again and again as the stool will re-grow. Sweet Chestnut and willow are good examples of coppice varieties, and can be cut every five years or so.

Felled trees can be left in situ to drop leaves and then be cut into rounds or manageable lengths for seasoning. Then if you have a ride-on tractor with a clip-on trailer or larger tractor just move and stack wood in manageable stacks near a path. The following winter you can collect and cut to appropriate lengths for the fire and then move closer to the house, stack and cover to season completely. If you have a large area of woods and don't mind the wholesale destruction, you can ask a logging contractor (*bûcheron*) to come and wholesale cut your timber, but be aware the woods will look devastated for at least a year.

Alongside your harvest of wood, you can also harvest mushrooms. Actually, you can do this whether you own woodland or not. Mushrooms can also spring up in pasture and it is possible to forage for mushrooms, but always ask the land owners permission.

Mushrooms appear all over France from April until November, depending on the area and the weather. There are over 3000 different varieties of fungi in France, including the truffle, but only a few are edible. Well; they are ALL edible, but some only once! Don't be one of the 1000 people annually in France who are diagnosed with fungi poisoning.

Arm yourself with a good identification book or better still, a knowledgeable local friend when you go hunting and take your haul to the local pharmacy, where trained staff will identify the safe from the poisonous. They do this for free, so there is no excuse not to. If after eating mushrooms you feel unwell, there are many anti-poison centres in France. Phone the *urgences* (15) and keep some of the suspect mushrooms or food for inspection.

Favourite mushrooms in France include girolles (in the UK we call these chanterelles), cèpes, morilles, bolets and of course, the truffle. Always use a sharp knife to cut and place in a wicker basket for collection. Always pick the whole mushroom, including any sac it emerges from (these ones with the sac tend to be the real nasty ones) as this helps identification.

The family known as boletus do contain some poisonous species so beware and ensure you have any specimens identified correctly. In the UK, the edible boletus(cèpe)are commonly called 'penny buns'.

The French believe that mushrooms appear a few days after a full moon, which coincides with rainfall. Check out chestnut woodlands and around oak and lime (tilleul) trees.

For more information on mushroom identification look at: **www.atlas-des-champignons.com**

VINEYARDS

Occasionally, a house with a vineyard will come on the market, and you may be tempted to consider this as a project. The lure of countless bottles of your own vintage in the cellar or cases of wine to sell is a pipe dream for most. Certainly, buying a property with a house and an established vineyard isn't going to be cheap. Running a vineyard takes years of experience and lots of hard work, so perhaps not the 'corking idea' you thought when it first jumped into your head. Conventional viticulture in France employs the use of a cocktail of chemicals and systemic fungicides and many vineyards have to display perimeter notices after spraying warning against access for up to 48 hours. Organic production is a fast growing market in France as French and foreign wine buyers are increasingly aware of the amount of residual chemicals in wine.

Wine-growing areas in France to consider if you are seriously looking to buy a vineyard or plant your own include Alsace, Bordeaux, Burgundy, Languedoc, Loire and Rhone, but many local wines are successfully grown outwith these areas. A Bordeaux appellation could cost you from €18,000 per hectare, whilst in Côtes de Provence, you'll need to spend at least €70,000.

Is the vineyard well-kept and virus free? Is there a winery on site, and if so is the equipment up-to-date and maintained? What grape varieties are being grown? These are just some of the questions you need to assure yourself about. Then you will have to research the wine industry regulations and find a good specialist lawyer and accountant.

Even if you avoid late frosts and heavy late summer storms and manage to get the harvest in and fermented, there is a long wait before you will know what kind of quality your wine is. Then the cost of bottling, packaging, marketing and selling is a full time job and success is not guaranteed. If you do indeed have a large expanse of grapes of a sought-after variety and quality, a good option may be to take them to the local co-operative to be processed, but please do research this before taking on a vineyard.

For some background reading on the subject of setting up in winemaking in France, please read Caro Feely's books on the subject. Her second book, *Saving Our Skins* is particularly relevant on examining what is involved in making wine commercially in France. Another good source of information is Alexander Hall's website: **www.vineyardintelligence. com**

PASTURE

Natural meadows and fields are often full of wild flowers which add to their charm but won't put money in the bank for you. Wild flower meadows often indicate poor soil (which is why the grass cannot out compete with the weeds), but offer a fantastic habitat for insects, especially butterflies. All fields and meadows require cutting at least once a year (*fauchage*), and the resulting crop, when dried and assuming it is free of poisonous plants such as Ragwort, make a useful commodity called hay. Many horse owners make a weekly walk around their paddocks to hand pull this weed and remove to be burnt.

Small bale hay is easily sold on to smallholders and horse owners, but again, at around €5 a bale once a year, isn't going to make you a millionaire. The process of knowing just when to cut, dry and bale is fraught with anxiety, especially if you have to rely on a baling contractor who has other customers in the queue. If you intend to keep other animals such as goats, sheep, cattle or horses, then the slurry or dung can be returned to the field which will enrich the soil and with time, perhaps allow for two cuts of hay.

On soil that drains freely you may find that you can keep livestock out nearly all year, but areas with poor drainage or high rainfall mean that animals, especially heavier animals will compact and poach the land, and tend to be housed indoors over winter, which of course means higher feed bills and increased cleaning.

If you really want to get as much value and enjoyment from your land as possible, then do think on having animals. The symbiotic relationship benefits the land, the animals and you, the owner. High value livestock, such as sheep will also make it financially worth the effort. After all, there are only so many eggs I can consume and the cost of production is pretty high for the return; but with lambs, you can literally make a killing after 5 or 6 months, and stock your freezer for the year to come. And, as I'm sure Kate Humble for one would agree, they are much more cuddly than chickens.

Some animals are choosy browsers; such as horses. Running horses in a field for a few months then removing to clean pasture is a good idea, but don't let the field stand empty – move some sheep or goats into it to follow the horses, and they will clear the grazing more evenly. Every time you move animals or poultry to a new field, worm them the night before. Then any worm eggs are left behind in the old field which can be left to recover for a while. Liming the field or spraying with dung or slurry, then harrowing will rejuvenate this area of land in time.

LAKES & PONDS

With modern sanitation and drinking water available at the turn of a tap, it's hard to imagine that as little as 30 years ago many people living in farms and hamlets in France were reliant on well water or springs to supply their drinking and cooking water.

Ponds and lakes (*étangs* and *lacs*) provided water for washing, laundry, watering of animals and activities such as manufacturing and blacksmithing, and many still exist on farms and in hamlets and villages. Large, deep ponds and lakes were often used to breed fish to supplement the food supply.

A catch-basin pond is usually man-made or an adapted pre-existing hollow fed by water run-off and rainwater. Beware of any nitrate run-off from fertilised fields. A spring or stream-fed pond is fed by underground springs or ground water which keeps the pond full and also provides for some water movement which in turn affects stagnation.

All lakes in France should be registered with the Department of Agriculture and fall into fishing categories (*catégorie piscicole*). Of course, this being France, some are unregistered, particularly if they are filled by rainfall and not fed by a running stream. If you are buying a property, ensure you or your *notaire* check. Registering a lake is theoretically straightforward and can be done via the *Direction Départmentale de l'Agriculture et de la Forêt* at the prefecture. For most purchasers the appropriate classification is category 2 (closed or *eaux closes*).

You can install a solar aeration system to prevent stagnation and help aerate a pond to keep wildlife and fish healthy. Other maintenance is required also. Ponds and lakes eventually silt up and need careful cleaning out and the soil and organic debris removed are great for potagers. Excess pond vegetation also needs removal. Be aware that draining down a lake requires official permission.

Small numbers of ducks and geese can use a pond without causing too much damage, but will puddle the banks and edges and defecate in the water. If you have the luxury of a stream entering and leaving the pond this will help to clean it and aerate it at the same time.

ORCHARDS

An orchard is a thing of beauty, and although orchards are viewed as traditionally British, there are many areas in France where they thrive. Normandy and Brittany and around the Basque region are famed for their cider and cider orchards, and in North Dordogne and Limousin there are many hectares of commercial dessert apple orchards. Most rural *fermettes* and houses in villages will have the odd apple tree or two or small mixed orchard, which is a wonderful asset. Alongside all the fantastic apple products you can make and enjoy, the orchard itself is a thing of beauty of usefulness. Whilst growing one crop, you can also undergraze it with poultry or ducks. Mature, traditional orchards could also cope with fattening lambs. Young individual trees will need serious protection from sheep, deer and wild boar in the way of triangular solid guards with chicken wire stretched over to a height of around nearly 2 metres. Electric fencing might be a consideration if you have more than a few young trees planted.

Current regulations from Animal and Plant Health in the UK allow for movement of fruit trees from the UK to France. If you fancy having some British varieties that you think will do well in your chosen area of France, then you are not obliged to buy French varieties. There is a demand, for instance, from British ex-pats for a good Bramley tree, as no French cooking variety can quite match a Bramley for taste and texture. The further south in the country you go, the harder it will be to find a good variety that will grow healthily and provide a crop. Concentrate on rootstocks that grow larger or medium sized trees, such as M25 or MM106, as these have more vigour to cope with demanding conditions and choose planting areas less likely to be affected by drought and wind damage. Spacing the trees further apart will eliminate competition. Varieties that mature earlier in the season will deal better with hot, dry conditions, but remember early varieties tend not to keep well, so utilise by freezing, juicing or turning into cider.

Of course, having a mixed orchard is a better prospect for a sustainable smallholding. A mix of apples (a cooker, and perhaps an early and mid season dessert apple), a dessert pear, a plum tree, a couple of cherries, and perhaps a peach or nectarine will have you collecting all sorts of fruit for pies, cake, jams and eating fresh. A fig tree is a good addition and some soft fruits such as currants or raspberries can also be sited on the edge of the orchard. Be realistic as to what you can do with all this fruit. A mature apple tree on M25 rootstock can produce a tonne of apples. Then you have to move them, process them and remove or process the waste. Unless you aim to commercially grow apples, a couple of trees will produce more than you could process.

For more information on orchards and cider making visit my Facebook page **www.facebook.com/SpottyDogCider.**

Cheat's Cassoulet recipe

2 tbsp olive oil or duck fat
250g chopped smoky bacon/ lardons
I sliced onion
3-4 chopped cloves of garlic
4 chicken thighs with the skin on (or duck or fatty pork)
3-4 Toulouse sausages
3 x 400g cans butterbeans (rinse & drain)
300 ml chicken or duck stock
2 tsps. tomato puree
400g can chopped tomatoes
2 bay leaves
2 tsp dried thyme

Turn the oven on to 170 degrees C. Heat the oil or duck fat in a frying pan and fry the bacon/lardons until soft. Tip these into the casserole. Add the onion and garlic to the pan and fry until soft then tip into the casserole. Fry the chicken (or pork or duck) in the oil until golden and put in the casserole. Fry the sausages until brown and put into casserole. Put the beans into the casserole. Pour stock into the fry pan, deglaze and bring to boil. Add the tomatoes and the tomato puree and the herbs. Stir well, tip into the casserole. Pop lid on and cook in oven for 30-40 minutes.

Chapter 4
Livestock

POULTRY

Having a few hens about the place is usually the first step into smallholding and an easy introduction to keeping any stock. They are inexpensive to buy and feed and can provide eggs for the table and add a pleasant 'farmyard' feel to any country cottage or house. If you are having just a few birds then no legislation is required. You can have up to 50 birds before you need to get a *cheptel* number (The French equivalent of a farm holding number in the UK).

They come in a standard size (large fowl) or a bantam (slightly smaller). A standard commercial point-of-lay pullet will give you an egg a day for around 18 months, then production will tail off. A bird from a recognised breed such as a Light Sussex, Cochin or Marans may produce less than this, but for a longer time. Hybrid hens tend to live shorter lives than traditional breeds; around 4 years compared to 7. Details on the egg production for breeds can easily be found on the internet, but on average four birds will produce 24 eggs weekly. As daylight dwindles in late autumn, egg production may also falter only to pick up again in spring.

If you also want to keep birds for meat you can choose an exclusively meat bird, such as the *Sasso,* which you will have to buy in, normally as chicks a few days old and which grow rapidly to eating size at between 90 and 140 days (roughly 20 weeks) depending on what they are fed and if housing has 24 hour lighting. Keeping cockerels from traditional heavy breeds such as Light Sussex or Orpingtons can also be fattened for eating, although they will mature at a slower rate and around 5 months is the optimum time for killing. If you wish to keep both egg producers and meat birds then keep them separated so that you can adjust the rate of feeding.

Poultry can be kept as barn birds or as free range birds. Barn birds normally are kept in lit indoor housing large enough to allow movement and with a nesting area or nest boxes for egg retrieval (if choosing an egg producing breed). Free range birds can be kept in a moveable or stationary poultry house with a door that opens either into a large pen or direct into a field or yard. Birds are let out in the morning and will return to the house at night for security. There are many versions of a hen house available to buy or build but you must take into account how many birds of what size when choosing the size of house. Always overestimate the space slightly. Choose houses which allow for easy cleaning and disinfection and allow good ventilation. The high temperature in French summer mean ventilation is very important, as most ill health in poultry is down to poor ventilation and dirty housing.

Also choose housing that is easy to move around. If you have to do this yourself, a large, heavy house is a real issue to heave around, but moving houses and runs will allow poultry access to clean, parasite free ground, which lets used ground recover, significantly deters rodent infestation and prevents smells. Mow the grass fairly short in runs and the sunlight will destroy any worm eggs and larvae in the soil.

Naturally, you want to choose young, healthy birds. Point of lay means the bird should be around 16-18 weeks old and come into lay at around 20-22 weeks. They should be fully feathered, bright eyed, be parasite free with clean eyes and legs. There should be no wheezing, sneezing or matter in the eyes and no swollen face. Pullets (young female birds) have a pink comb. Young cockerels are usually taller, not as plump and may have a redder, more pronounced comb and also red wattles (skin folds under the beak). They also have sickle shaped neck and saddle feathers compared to rounder feathers on the females. You do NOT need a cockerel to get eggs from your hens. You only need a cockerel for breeding.

Local prices will vary, but unless you are going for a very rare breed you shouldn't be paying more than €35 per pullet. Places to buy from include markets and farm stores who often have signs up with details of birds available locally. There is a good Facebook group called **Poultry Keepers in France** with many members to give information, advice and sell stock.

Birds need letting out daily in the morning. The house can then be checked and any eggs collected. The sniff test will tell you if the birds need to be cleaned out, but generally once weekly is the average for complete emptying of the house, sweeping thoroughly and replacing the floor litter. Wood shavings are a good medium as it absorbs moisture and does not mark eggs. Avoid using chopped hay or straw in the coop as the birds will soil and wet this and soon it will be a source of fungal spores and the birds may show signs of illness such as wheezing. Wood shavings are perfectly adequate.

Birds should be fed and watered outside (unless they are barn-kept birds). Chickens need a lot of water, so ensure you have clean, fresh water daily. In the summer a single bird can drink up to 2 litres of water in a day. Food can be given in trays, feeders or on the ground. I liked to fix plastic rainwater guttering along a fence about a foot off the ground. The birds could reach in to feed but there was little wasted food and it did not encourage rats or mice.

Food should be given twice a day, and free range birds will supplement this by 'free-ranging' on beetles, insects and worms on the ground they have access to. Some larger birds are known to kill the small snakes that exist around old buildings. If they are free-ranging birds you need not add grit or oyster shell to their feed unless they look as if they require it during moulting time. Keep chicken feed inside a damp-free building inside metal or heavy plastic bins with tight fitting lids. Rodents getting into bird feed will foul it with their urine and droppings, and mice and rats carry salmonella and other horrible diseases.

When cleaning out housing you can compost the used material, which will take around 6 months to break down. Mixing in vegetable or grass waste will aid the process. Check housing regularly for red mite, especially in the summer. These tiny creatures live in small crevasses in the house during the day and return to feast on the birds at night, sucking their blood. A piece of white kitchen roll wiped underneath the ends of perches will reveal any. Blood smears mean you will have to empty and disinfect the house with a proprietary red mite killer, such as Mite Kill, Ficam W or, if you can't get these use diluted Jeyes Fluid (wearing appropriate gloves and old clothing). Treating the house again in 5 days should catch any newly hatched mites. Birds can be treated with Frontline or similar, (which will also deal with any chicken lice) but be sure to observe the egg withdrawal period and use the correct dosage. In the UK, I used 3 drops per pullet and 4 per cockerel or large hen. Push back the feathers to reveal skin and drop the drops direct onto the skin. This should alleviate the problem for a few weeks or months.

Placing a suitably sized cardboard box inside the house as a nest box will also provide a refuge for any mites to hide in during the day and can be removed and burnt periodically.

Worming quarterly will also ensure your birds are healthy and laying well. The only licensed wormer is Flubenvet, which can be added to pellets or mash with a little sunflower oil to ensure the medication sticks to the feed evenly. Worming will kill worms in the gut and also gapeworm in the throat. As a pick-me-up after moulting or if a little under the weather, apple cider vinegar added to the water seems to improve bird wellbeing.

Hopefully, all this talk about worms and lice etc hasn't put you off keeping poultry or animals completely. Well kept, clean birds rarely have problems, but pre-warned is pre-armed. Chickens can present with illnesses, but Marek's Disease is rare and Avian Influenza extremely unlucky. Marek's presents suddenly in young birds approaching 18-20 weeks old. Birds suddenly go off their legs or drag one wing down, often with a tilt to the head or neck. Culling is the only option and it is highly infectious. Most commercial hybrid birds are vaccinated against the common poultry diseases, and as such are a good choice for the novice hen keeper.

Avian Influenza (Bird Flu) is brought in by wild migrating birds and affects all poultry and waterfowl. If you live in an area where migrating birds visit or pass, you need to be aware of any outbreaks and be prepared to contain your birds and report any suspicious deaths to your vet.

You may decide after a little while that you'd like the idea of chicks around the place. This either means getting a cockerel (either full time or just visiting for two weeks to fertilise the hens) or buying in hatching eggs. You can hatch either under a broody hen or with an incubator. I've tried both and found some broody hens useless and some great, but I have to admit to liking the incubator and then raising the chicks myself. The issue here is you will either have to buy or borrow an incubator.

From fertilisation and the egg being laid until hatching is 21 days for large fowl poultry. There is plenty of information on the internet and from incubator manufacturers.

Humane Dispatch

Of course where you have life you also have death, and knowing about humane dispatch is something every smallholder needs to know. Even if you never intend to eat your own animals or poultry you need to know what to do in an emergency. This emergency may be the visit from a fox, badger, weasel or raptor, which leaves your bird not dead, but badly injured or very shocked. The emergency may be the gradual onset of old age or illness or a sudden deterioration.

You have 4 choices. Ignore it; which means you seriously need to examine if you are a fit person to be keeping livestock without attending to their basic needs. Take it to the vet to treat or euthanase; you've done the right thing, but it may be expensive and if the emergency really has been urgent the bird may die in transit. Get someone locally to do the deed for you. Or learn to do it yourself in the quickest and least stressful method to you and the birds.

I've tried the 'wring the neck at arms length' method. I can't physically do it. As a woman my arms simply aren't long enough. So, I have learned the broomstick method. It's foolproof, and once learned will deliver a quick, clean death. It's also a good method for those who are squeamish as it offers a degree of separation between the human and the bird.

First of all, don't use a broomstick – it's far too large in diameter to do the job cleanly. Find or buy metre long length of 12 or 13 mm round steel bar. This will be your broomstick.

The method is to remove the bird from the others (easier when they are in their house), holding it securely and taking it to a quiet area. Place the bird facing down on the ground whilst keeping tight hold of the legs. Place the bar over the back of the neck. You will know you have the right place as the bar will neatly sit in the little dent between the neck and the back of the skull. So you now have a 'cross' made of the bar lying at right angles to the head and body of the bird. Use both hands to firmly grip the legs as far up the bird as you can comfortably.

Place one foot firmly on the bar about a foot from the head. Repeat with the other foot on the other side. Pull sharply and firmly up and forward in a smooth but forceful pull. You should feel the neck go instantly, but if it's your first time you will be nervous and may not notice. How hard to pull is the question I'm always asked. This is hard to answer as we all have different ideas of what 'sharply and firmly' is. Practice will perfect your method, but if you are pulling the chickens head off then you are pulling too firmly.

The head dislocates from the neck, pulling the spinal cord out of the base of the skull rendering death very quickly and painlessly.

The bird will start to flap (keep a good hold of the legs) and this may take a few minutes to subside (less if the bird was near death). This is muscle reaction only. The bird has been dead since you pulled the bird up and forward with the bar over its neck.

PRODUCTION FOR THE TABLE

If you have been producing your birds for your own consumption or feeding excess cockerels that you have hatched yourself, then naturally you will want to prepare them for the table. We have found that restricting a cockerel in a small pen reduces a little the muscle build up in the legs, and supplementing the standard feed with a feed of mixed corn or cold cooked pasta in the evening adds to building up the weight in the bird. Killing at around 4-5 months will produce a decent bird.

If you have chosen a commercial meat bird, such as a *Sasso* then you will have birds ready for the table a lot sooner as these commercial birds are bred for fast finishing. Ensure that any bird you choose to eat is healthy. If it is an old bird it will be tougher than a younger bird; if it's an egg laying breed then the effort of preparing for the table probably isn't worth it.

Before your chosen birds have been culled, prepare yourself and your working area carefully. You will need sharp knives, buckets, refuse sacks, a place to suspend a bird so that it is at chest height and a sturdy table covered with washable plastic. An airy but windproof shed is ideal. Prepare a couple of buckets of warm water, some soap and a hand towel, some large freezer bags or similar for your finished meat and arrange to do this on a quiet, cool day when you will not be distracted.

Take the bird and tie or band the legs together with string or baler twine, and hang head down in the shed from a hook suspended to allow you free movement. Start plucking the long wing feathers first whilst the birds is still warm (the feathers come out a lot easier). Primary and secondary feathers usually one at a time (you may need pliers to grip them), then small fingers full of the smaller softer feathers. Once both wings are plucked, start on the body, working your way down the bird. Pulling a few feathers at a time reduces the risk of tearing the skin. Pull in the opposite direction of growth. So, if the bird is hanging upside down by its legs you should be pulling the feathers down the way until you end up at the neck. Leave the bird hanging a few minutes whilst you sweep the feathers away.

Return to the bird and remove from the hook. Take to the table and arrange a large bucket with a bin liner fitted inside to be handy for all refuse. With the bird at right angles to you, sever the neck and remove the head into the bucket. Enlarge the hole in the neck slightly, where the neck and breast meet and you will be able to reach inside this small opening with a clean tea-towel to grasp the windpipe and pull it towards you. Cast this into the bin.

Turn the bird face down on the table with its legs near you. You will see the vent area which, with a small sharp knife you cut round without cutting into the intestine. Gently reach inside with your fingers loosen the intestines which should fall out of the hole with a little help into the bucket. When you think you have all the intestines out feel inside for the ribcage. The lungs, which are almost impossible to feel, will be situated on either side clinging to the inside of the ribs. Scrape these away gently with your fingers. Remove the tips of the wings with pliers or butchers scissors if you wish.

Take the bird and rinse inside with clean cool running water. Try to keep the outside of the bird dry. You can then sit the bird on its belly and 'arrange' it to look somewhat like a supermarket bird. Pop it into a freezer bag, date it and then allow to cool completely before placing in the freezer. After processing the first couple of birds, you will become quite adept at this and the others will be a breeze.

If you think this is a long process, then you can simply joint the bird. You will lose a percentage of the meat, but by simply removing the skin and feathers like a glove, and removing the breast fillets and legs, you save yourself a whole lot of time. Again, wash the fillets and legs in cool, fresh water before bagging & freezing.

This is the same process for dealing with pheasant or duck, although if you pluck ducks, there are a whole lot more feathers, but you can get them ready for the table at 14 weeks, so one incubator load of 24 duck eggs should give you enough duck to provide for at least 18 or so roasts or a lot of leg and breasts within just a short time.

DUCKS

Ducks are great fun and worth considering if you have any natural water on your land. They can be messy in wet winters and will foul their water with their droppings, so natural ponds with water in and out or small streams or rivers are ideal to refresh the water. They need a suitable secure house with a large door and enough water to enable them to fully immerse their heads. And of course if you have ducks of both sexes you will be able to have ducklings and be able to produce your own ducks for meat, should you wish. One drake needs at least 3 or 4 ducks to keep him happy.

There are many types, colours and sizes. If you want just eggs and don't intend to eat duck, then consider call ducks; small, pretty with a loud quack, runner ducks; taller with an upright stance or Campbells, which come in khaki or white.

Dual purpose egg and meat birds include Silver Appleyards and Muscovy ducks. Silver Appleyards come in a large and small variety and are undoubtedly pretty. Muscovy ducks have the unusual red facial masque and are very like geese in habit, grazing a lot and they also fly well, so may need wing clipping. If you have a Muscovy drake with other duck breed females and you have ducklings from this mating these will be sterile. The ducks make excellent broodies.

For meat any of the dual purpose birds above, Rouen's or Pekin's or Pekin crosses make good meat birds. Rouen's are big, blocky birds, excellent for meat but who lay pretty well too.

If you have no water you can use a small baby bath or plastic shallow container set into or on top of the ground. This should be easy to empty and clean and refill. The water can be used to water plants nearby. You will have to 'train' both ducks and geese where their home is, so when you first get them pen them in for at least 2 weeks to imprint on them where home is. You will have to drive them into the secure house at night, which may require two people, but they will learn if you persevere.

Essentially, they eat the same as poultry, although are greedy feeders, so keep separate from poultry and allow the hens out for feed first. Ducks will lay eggs inside the house if you keep them in for at least an hour after daybreak. You may have to wash the eggs (warm running water) as they are dirty birds in their houses and stubbornly won't use nest boxes.

Most ducks start to lay in early spring from around 24 weeks of age, and if you wish easy retrieval of eggs keep the birds in their house until after 9am, when they should have laid. You must ensure the duck house is kept clean and dry inside or the eggs will be filthy. A duck will lay well for up to three years but may live until 10 years old.

If you plan on raising a few to eat, then they should be ready at around 12-14 weeks, which is also around the time you can sex the birds. Drakes tend to show a little curl in the tail feather at this stage. Fatten ducks with pellets, grain, cooked cooled pasta and vegetable scraps.

To kill and prepare for the table is the same as per poultry, although the plucking is more laborious. You do not need to 'hang' a duck to improve its taste. If you want to rig up a cheesecloth cage to suspend the bird in, hang in a cool place and ensure that is big enough to allow space to prevent the bird touching the cheesecloth. The cage prevents flies laying eggs on the carcass.

GEESE

Geese are large birds which need a large area of grazing and a secure house at night. Don't underestimate how much grass geese will eat; 80% of their diet is grass, and supplementing their grass with a handful of pellets in the morning and corn in the evening, especially in winter will not do any harm. A quarter of an acre pasture can support five adult geese. If you intend to eat them, then obviously pellets and corn will fatten a bird much quicker than grass will. They also enjoy vegetables and some fruit.

Geese kept in a young orchard will strip the bark from trees and kill them. They make excellent property guardians and will honk and hiss when anyone or any car approaches. Toulouse geese are the preferred breed here in France, although you may also be able to source Embden's locally. I have been told that Toulouse geese are more docile than the Embden crosses we had in Cornwall, which were vicious birds. Geese have a serrated beak and a bite can break the skin and will bruise and hurt. They are great as guard geese, but sadly there is no guarantee that the fox will not take them.

Geese start to lay eggs in mid February and the gander may express a preference for one female and ignore the rest! If you intend to hatch the eggs bear in mind the first couple are usually infertile. Geese lay every other day so it will take a few days to get a clutch. Bear in mind that geese breed better in their second year, and again laying is good for three or four years, although geese have long lives and may well live until they are 20 years old or more.

If you intend to keep any poultry or waterfowl you must bear in mind the risk of Avian Influenza, especially at migration times. You must be able to secure your birds inside in secure housing and covered runs in case of a breakout near you. If you are aware of an outbreak you need to contact your vet to discuss what is required by law.

BEES

Bees are a relatively easy creature to house on your land or in a small garden and are not too demanding of time or effort. You don't need to let them out, feed them or put them in at night and by and large they take care of themselves with a little help.

There hasn't been a day gone by in the last five years when I haven't heard or read about the plight of the honey bee and how they are suffering these days. If you want to do your bit to support honeybees but just can't or don't want to have your own bees, then simply plant more daisy type flowers, especially if they flower in mid summer, which is a traditional time of few flowers being available. Leave some ivy growing somewhere and plant a few winter flowering plants such as Mahonia and heather for them to collect nectar on sunny winter days.

If you fancy doing more than this, then read on, because there are two ways to keep bees. The easiest and cheapest is to visit your neighbour or *mairie*'s office and ask who the local beekeeper is, contact him or her and offer them a site for one or two hives on your land in return for a jar or two of honey annually. The beekeeper (*apiculteur*) will do all the work, be responsible for all the regulatory paperwork and you can relax and know that you have done your bit for the local bee population.

If you want to take up beekeeping (*l'apiculture*) as a hobby; and it is a fascinating and rewarding one; then you need to start with some PPE. An all-in-one suit with built-in hood with mesh front is an excellent purchase. Ensure you get one big enough to be able to move and bend freely. They come in all sorts of colours but white is great as it reflects the sun and is therefore marginally cooler to wear. You will also need gloves or gauntlets, a smoker, a hive tool and some sort of hive. Join a club or find someone to mentor you in the first year or so until you gain confidence and knowledge and you won't go wrong. Having some anti-histamine tablets in your suit pocket is a wise precaution in case you get stung. If you have or develop an allergy to bee stings then speak to your pharmacist about getting an Epi-pen.

In the UK, there are many variants of hive. I myself used second hand Smith hives. They were cheap to buy and being slightly smaller and lighter than a National, were a good choice for me as I have back problems. In France the main ones available to buy are Dadant hives; which are similar to commercial hives, but equipment is refreshingly reasonable to purchase and easy to find in local agricultural cooperatives. Expect to pay around 200 euros for a complete hive (*ruche*) with frames, roof, super and colony of bees.

All hives are subject to regulation in France. The beekeeper must register annually in December at both the *mairie* and the Chamber of Agriculture. Colonies must be insured and are subject to an annual inspection. For details on regulations look here: Bee Keeping regulations France **ruche.ooreka.fr/ comprendre/reglementation-ruche**

Siting a hive away from the flow of people and animals is important for your own peace of mind and safety. Other livestock don't really mix with bees, and children and dogs need to be kept away from hives. You only need a metre square per hive, which can be sited in an overgrown area, at the edge of woodlands or even on a flat roof if you are short of space. Face the entrance of the hive East to get the sun in the morning. Shelter from the wind is important, but bees are pretty resilient and will, as a colony 'shiver' to keep warm on cold, wet days and enter a type of torpor in winter.

In a good summer the colony numbers build quickly from around 20,000 to 70,000 in midsummer and an average hive can produce 60 lbs of honey in a good summer. Taking off the harvest is done in August and then the bees are given a varroa treatment. All honeybees in France and in the UK now have been exposed to the varroa mite and the colony will appreciate the diligent beekeeper's attempts to reduce the mite numbers in a hive with regular treatments. A good colony will continue to collect nectar and make honey after the official harvest to store for their winter food.

The spread of the Asian hornet (Vespa velutina) in France has been well publicised and if you have them you are supposed to alert your *mairie*. They can be identified as having a dark brown body with an orange/yellow face. I can tell you in this wooded area of South Dordogne we have them in just as large numbers as the European hornet, and the summer of 2019 has recorded nests and sightings as far north as the Midlands in the UK. The issue with hornets is that they often predate on bee larva by entering weaker colonies that are unable to defend entrances and then eat and carry off larvae to feed their own young. This weakens the struggling bee colony further. Some manufacturers have been devising special hornet proof entrance blocks for hives, and removing any landing boards attached to hive fronts helps a little. If you have the misfortune to be stung by a hornet, take an anti-histamine tablet and apply ice to the stung area.

More information on beekeeping in France can be found on Chris Luck's site: **www.wild-life-in-france.blogspot.com/p/be.html** and also on **www.facebook.com/groups/BeeksinFrance/**. The late Dave Cushman developed a wonderful website devoted to bees which is still worth having a look at. You can find it at: **www.dave-cushman.net/**

SIMPLE HORNET TRAP

Take an empty plastic water or lemonade bottle and cut it in two about a third of the way down from the top. Remove and recycle the screw top. Invert the top part and push inside the bottom part ensuring a tight fit. Make two opposing holes for hanging. Fill with a mix of fish or meat, water and a little old beer or wine to a level just under the protruding bottle neck. This will attract Asian hornets which will drown in the liquid. Do NOT use honey.

RABBITS

Quiet and clean and taking up a small amount of space, rabbits can provide a high protein meat for the table with a little effort. You will need one adult male (6 months old) to a maximum of 5 female rabbits, but to be honest a female adult (4 months old) is capable of producing 6 kits (babies) every 45 days, so 2 females will be more than enough for the average household. You will need to replace your breeders between one and two years old, and remember that individuals from the same litter cannot breed together.

Buying your rabbits couldn't be easier, with many French towns offering live rabbits for sale in street markets or at agricultural stores. Food is generally rabbit pellets, but supplement with hay and fresh grass. If keeping on/in runs outdoors, also provide pellets. Fresh clean water is essential.

If the mother rabbit has a litter of 3 weeks old then she is ready to breed again. Take her to the male. Allow mating to take place two or three times then remove and return the female rabbit to her cage. Wean the kits at 5 weeks old and remove from mum and feed until they are ready for processing at 11 weeks or 5lbs live weight. New babies are born 28 -32 days after mating. Provide a hay filled nesting box for the mother at day 25. You need to handle babies from the second day to get them calm and used to being handled by humans. This should give you a carcass of around two and a half pounds.

You can either keep indoors in a purpose built shed with staggered cages along one wall, or keep breeders indoors and growers in movable secure runs on grass outdoors. There is plenty of information on keeping rabbits for meat and dispatch/processing on the internet and on YouTube. A series of helpful videos ends with this one on processing: Youtube Keeping Rabbits **http://www.youtube.com/ watch?v=aHmIHjCIRAY**

One important thing to remember is to keep rabbits clean, especially young rabbits. Coccidiosis is a horrible disease and can quickly kill rabbits. Keeping them clean and feeding some proprietary feed with a coccidiostat in it should prevent this.

GOATS

Goats are an option if you are planning on having home produced milk, cheese and meat. The most popular breeds in France are Saanen, Toggenburg and Alpine. You will need housing, (goats aren't waterproof and don't like rain or wind) and a securely fenced paddock. Bringing them in to be housed in winter is strongly advised, at least for the foulest weather; so you'll need to consider an adequate supply of straw for bedding and hay for eating, as well as pellets and green fodder. Goats can be the Harry Houdini's of the animal world, so high and secure fencing is essential. If you wish to milk them, a small dairy area is a good plan too; especially with a concrete floor you can hose down afterwards. Milking should be done morning and evening, throwing away the first few squirts which may contain bacteria. A stainless steel bucket is perfect. Milk left to settle will see the cream rise to the top. Goats' milk can be made into both soft and hard cheeses. If you wish to make your own, I've included a simple goats' cheese recipe at the end of this chapter.

Goats are browsers rather than grazers and appear to like stony, rough ground and are ideal for rough slopes. They will require more land than sheep, so 4 animals per acre in France is a good figure. They will destroy an orchard; reaching up on their back legs to strip any branches in reach, but are useful, as are sheep, to follow cattle in pasture.

Mothers will produce milk for at least 18 months after kidding, so if crossed with a Boer sire, you could produce kids for meat and get milk, and therefore goats make a useful animal for a smallholding. Breeding time is in autumn and pregnancy lasts 150 days. I'd suggest taking the females to a buck rather than buying one as male goats are really smelly. When the kids are born, allow the mother to feed them for at least five days, then separate the males who can be fattened for meat. Keep an eye on the mother to ensure both sides of the udder are being emptied to avoid mastitis. Strip out any milk if the udder feels hot or hard and consider calling out the vet if it does not improve after 24 hours.

SHEEP

Sheep are kept in relatively small numbers on French farms in the warmer regions. As the summers are quite hot from Dordogne southwards, the grass stops growing and in places dies off after July, thus giving grazing livestock a shortage of feed and owners the cost of supplementing this. The added burden of carrying wool means that traditional French breeds tend to be more favoured than traditional British breeds which have been bred as a dual purpose meat and wool breed. Many French breeds are also used for milking to produce cheese, and are taller to accommodate milking.

I like sheep and decided quite quickly after arriving here that our empty paddock needed some sheep. I missed the ones we had in the UK terribly, and the chore of mowing all the grass seemed like we were wasting an opportunity. We wanted to buy in young lambs to keep over the spring and summer, but couldn't find cade (bottle feeding lambs) locally and finding local farmers willing to sell us weaned lambs at a reasonable price proved very difficult. We got in touch with some ex-pat English smallholders who sell us a couple of weaned lambs annually.

Keeping lambs from just weaned through the early summer until July (or whenever your grass dies off) means you don't have to worry about shearing, winter feeding or breeding. You keep the lambs until they are judged ready for slaughter and off they go, or in our case, the butcher comes to us, and then we have beautiful home produced meat for the freezer. On our one acre field we can keep 3 big lambs from February until the start of July. We like Causses sheep, a hardy breed of sheep from The Lot.

Of course, some people are opposed to keeping animals for meat, but they are still very useful as lawnmowers and they will naturally keep the grass in check without the need for a petrol driven machine. Small, dark fleeced Ouessants are ideal for this. They are quite small and very hardy. They will need to be wormed, vaccinated & fed over winter and shorn in summer to prevent overheating and flystrike. We provide a shed for shelter from the sun, but if you had mature trees this would provide ample shade. If you intend to keep a small flock all year then a shed will double up as a lambing pen if you size it and plan the interior. The Facebook page: **Sheep Keepers in France** has some experienced keepers who offer advice and animals for sale.

PIGS

In the UK, the pig is a traditional smallholders' animal. They were easy to keep and ate leftovers and could be turned out in oak and beech woodlands to forage. This traditional lifestyle for pigs isn't really feasible now with a shortage of common land and strict regulations regarding what you can and cannot feed to any animals.

Pork is one of the favourite meats available to buy in France and much is made from it. Every part of the pig is eaten except the oink! As a result many small farms have pigs for home use, and so if you fancy having a couple, it's relatively easy. Pigs are intelligent and social animals and you must keep at least two to allow for them to live naturally.

You will need a decent movable pig ark or other housing, because pigs like clean, dry housing. Electric fencing is a great idea to contain them, at 6 and 12 inch height from the ground. These intelligent animals will quickly find any weakness in traditional fencing and be off into the nearest vegetable garden or crop field. There is no such thing as a non-digging pig, no matter what anyone will tell you about certain breeds. They all dig. You can turn them into an orchard when the apples are fallen, but you must keep an eye on them or your orchard will be turned into a ploughed field which will seriously disturb the roots of the trees.

They are natural woodland creatures and will enjoy any woodland you can offer them. Use their natural clearing ability to clear scrubby, overgrown bramble patches, and watch them manure it as they clear it. They need proper pig nuts or pellets and will also like vegetable waste and apples from the orchard; but as a supplement, not as their main food. Pomace from apple pressings can, if fresh, be incorporated into pig (and indeed sheep and cattle) rations. Pigs will grow quickly and soon need a new enclosure if the existing one is too small or becomes muddy and dirty. Oh, another fable is the micro pig. All piglets are small, and they all grow quickly into large animals.

So, if you fancy getting a couple of weaners of the same sex you will, in around four months have animals ready to go to the butcher. Decide early on whether you want to produce for pork or bacon, and chat to your local butcher if he will produce sausages for you and at what cost. A 6 month old porker will produce around 45kg of meat, so you need to think how you will use the meat from two!

Have a look at this Facebook page for more advice & information: facebook group **Pigs in France**

CATTLE

Although most smallholders won't really want to consider having cattle, we will examine it here. We have already spoken about the hot summers and the grass being insufficient for feeding smaller animals like sheep and goats, so obviously, if you are thinking of having cattle you will need a serious amount of land to rotate grazing for them. Geography will also dictate what breeds of cattle are suited locally. A highland cow is not going to cope with sweltering summers of 40 degree heat. Buying cattle locally not only provides you with acclimatised animals, it also lets you build a rapport with the vendor who could be a good source of advice. Finally, a cow is a large animal. If you struggle moving or examining sheep, then you really need to consider if you can manage a cow or cattle. Physically, you'll need to be fit, especially if you are considering having an in-calf cow and need to be able to assist in a difficult birth.

Dexter's are an ideal smallholders' cow. They are quite small in stature and can be reasonably biddable. A cow can provide you with milk, cream, butter and cheese, and if you have a cow with a calf to raise you will also be able to provide meat. A Dexter killed at around 20-24 months old will provide around 150-220 kg of meat. Remember that a milking cow needs to be milked twice daily and if you have to be away for any time, finding someone to do this won't be easy. There are some trusted house sitters and farm sitters who will do this for you, but the service isn't cheap and good sitters are booked a year in advance.

You'll need to provide at least one acre of decent pasture per animal with shelter in winter. A barn of around 4 x 10 metres is sufficient for two animals. You will have to muck out daily to prevent illness and flies, and composting the material will provide great nutrients and soil in the potager. Remember to provide fields with clean drinking water. Cows can drink 30 litres plus in a day.

Removing the animal from the field to a shelter or barn prevents poaching of the soil. In France this usually starts in December and cows can be returned to the field in April. A further acre can be used to provide hay for supplementing feed in winter, and rotating this hay field with the pasture field will help keep the land clean and parasite free. Growing your own hay will provide feed and save money. At least half a small bale of hay per animal daily in winter soon mounts up.

HORSES, DONKEYS & ALPACAS.

Not strictly smallholding stock, horses and donkeys are frequently kept on smallholdings or houses with fields. Far too often donkeys are neglected and can be seen offered on social media or locally. Remember both horses and donkeys need regular visits by a farrier to keep their feet in order, as well as the usual requirements that animals need. You may need to call out a vet to deal with troublesome teeth or other issues, so be prepared to spend regularly to look after these animals. All equidae require horse passports, microchips and if you keep 3 or more then will require vaccinations for flu and tetanus also. Check animals all have a Carte D'Immatriculation issued by Haras Nationaux (this is paperwork proving ownership). If you are interested in giving a donkey a forever home look at the Case Study for Hazel Rudge, who runs a donkey sanctuary in North Dordogne.

The IFCE (*Institut français du cheval et de l'équitation*) regulates all equidae in France. For more information you can visit: Equine movement regulations France **www. equinerescuefrance.org/regulations-laws/passports/**

Llamas and Alpacas are a whole different kettle of fish. They are neither pets or classed as farm animals (probably because they hardly ever enter the food chain). No pet passport is required for them, however they must be microchipped and have a Certificate from ESIRECAM who hold the central database for Llamas and Alpacas in France.

In the UK, llamas and alpacas are seen more as a smallholder 'pet' rather than a farm animal. Generally, they are not eaten, and there is a limited 'crop' of fleece from them. However, they are popular, both in the UK and in France, where enterprising farms and holiday accommodation providers offer 'llama trekking experiences'.

They are expensive to buy and of course are a herd animal, so you will need at least two. The stocking rate is around 3-4 per good acre in France and they will require shearing once a year, by an experienced shearer. They can produce 3-5kg of fibre and benefit from shelter in the winter and from summer heat.

It is common to see them in a field together with sheep or

poultry and it is said that in their native South America they will chase foxes who are known to take the babies (cria), and as such are fox deterents.

BUYING & TRANSPORTING LIVESTOCK

The rules and regulations for transporting animals in France are the same as the ones used in the UK (pre-Brexit), so although the paperwork will be in a different language the rules will be the same. Trailers must comply with the EU regulations for the correct transportation of animals, you must complete an animal transportation course unless you are using a recognised and registered animal carrier to transport them. Animals must be accompanied with movement licenses, one of which goes with the transporter.

Moving animals TO France from the UK will probably get a lot more complicated once the transition period has passed after the U.K. leaves the E.U. At this time, export papers must be completed, and all veterinary testing and certification must be done by ITAHC approved vets. You cannot move them via the Channel Tunnel. You can find up-to-date information from your vet or search for **Defra official vets guidance sheep goats**

ABATTOIRES & HOME SLAUGHTER

If you wish to keep animals for meat, then you will have to consider how and where you are going to slaughter them. EC regulation No 1099/2009 On the Protection of Animals at the time of Killing will give you the basic information you require in the form of factsheets; which make for fairly graphic reading. Ec.europa.eu/food/animals/welfare/practice/slaughter/2018-factsheets-en

Stunning is required unless the animal is being killed in

accordance with religious practices. You are not allowed to home slaughter cattle, horses or donkeys. Pigs, sheep and goats CAN be slaughtered at home, but officially this slaughter must be declared before the municipal administration. Home slaughter of poultry and rabbits (where the meat is intended for consumption by the immediate family and NOT for sale) is allowed and must not be declared to the municipal administration. A private slaughter is the slaughter of any animal whose meat is exclusively intended for the consumption by the owner and his immediate family. This private slaughter may take place at home or in a slaughterhouse (abattoir). More information can be found here: **www.afsca.be/rapportact.../2018/ inspections/abattage/**

There are few abattoires within France, and some will take only specific animals. For example, in Dordogne (24) there are currently 4, but in Oise (66) there are none, so be prepared to travel if you wish to arrange your livestock to be processed at an abattoir. From experience, I would ask your local farmers what the arrangements are for processing livestock.

DEATH & DISPOSAL OF LIVESTOCK

Finally, as a smallholder, you need to keep your animals responsibly. Keeping a clean and tidy smallholding will deter rats and mice and the diseases these carry; you will have healthier animals and less chance of nosey or jealous neighbours complaining about your lifestyle.

Clean water is essential for all animals and birds. Do not use water from water butts as this contains parasites and especially in young animals and birds can spread coccidiosis.

There is a great Facebook page called **Responsible**

Smallholders in France with lots of information and contacts.

If your animals appear to be ill, you have a duty of care to seek a vet to prevent them suffering. If you have to euthanise an animal, remove it from the others and do so calmly and with as little stress to the animal or yourself. If you are unsure or baulk at this, get someone else to do it or call in the vet. Sadly, through either age, illness or incident animals die. You can no longer bury dead farm animals on your land and must dispose of them legally. Whilst the family dog or cat can safely be buried under it's favourite tree (very deeply to prevent wild foragers digging it up again), you cannot do this with a sheep or goat.

The procedure is that you contact your local departmental *équarrissage*, who will arrange for the French equivalent of the knacker's van to arrive at your property for collection and disposal. The system is now all online and price is according to the size and number of the animals. After arranging and paying, you must place your carcass on a pallet and leave outside your property entrance for collection. Covering with a tarp may do nothing for the smell, but most departments collect within a day of being notified. They will attend and check the ear tags of the animal. They will not take an untagged animal.

Bondon (Goats) cheese recipe.

2 litres (4 pints) fresh goats milk
125 ml (1/4 pint) cultured buttermilk*
3 drops rennet

cultured buttermilk is made from skimmed or semi-skimmed pasteurised milk which has been inoculated with a lactose producing starter. This is the buttermilk you will find for sale in supermarkets.

Mix the milk and buttermilk together, heat to 18C degrees then add the rennet. Leave overnight to curdle and then drain and ladle the curds into a cheesecloth. Hang over a bucket for 2 hours to drain, then wrap in a fresh cloth and press overnight with a clean heavy wooden board with a weight on top. Unwrap, sprinkle completely with salt on all surfaces and mould into cylinder shapes. This soft cheese will not keep but is worth making.

Chapter 5
Grow Your Own

Living the Good Life doesn't just mean keeping animals and eating animal products. There is a whole other world of fruit, nuts and vegetables out there. You'll be able to fill your freezer and live a comfortable existence with healthy, home grown produce, home brewed beer, cider, wine and liqueurs......

HANG ON! When are you going to find the time to do all this?

Well, time management, physical health and ability are important considerations, so let's look at what you really think is possible and what might be best avoided.

When we moved to France from our smallholding in Cornwall, I was adamant that we were stopping doing all that self-sufficiency stuff. No more animals to tie us down, no vegetables to weed, no country bumpkin stuff. My son quite rightly didn't believe me, and should have put a bet on just how long it would be.

When we moved I'd brought 15 British apples trees to France with me, and all my juice pressing and cider making equipment; so I could continue to make juice and cider for ourselves; my excuse being that an orchard is a thing of beauty and incredibly useful and that looking after it wasn't really work.

I also brought some demijohns as I do enjoy a little tipple of home-made wine. John brought his anvil, 2 welders and all his tools as we would need to be prepared (he said) for anything. We arrived in the September and by January I was looking for someone to sell me lambs as I felt bereft without them. Sadly, farming and rural living is part of me, but we have made significant changes from what we did in Cornwall. We don't have any poultry and it's not on the horizon. For the time and effort and cost we can buy great eggs, chicken meat & duck meat here. We therefore have no birds to tie us into getting back before dark to secure them, and no laborious housing to clean out. Our lambs stay with us for a maximum of 5 months; so it's not hard work making sure fencing (*la clôture*) is secure and that they have clean water and are happy and healthy.

Our woodlands need work in the winter and early spring, but we actually love having the woods! We see lots of wildlife, including deer, red squirrels, a multitude of birds and butterflies and the carpet of wildflowers is getting better as we open up the canopy a little.

Fruit and vegetable production is the sticking point. All our neighbours have *potagers,* and these provide vast quantities of vegetables and salad crops. However, they work on them every weekend, and I've been reluctant to start one until now.

Meantime, fruit production is a subject close to my heart. I LOVE fruit! All sorts of fruit, and our climate here in south Dordogne means we can grow all sorts, which is great. And most fruit is really easy to grow and doesn't take up much time. So, in this chapter I'm going to share some ideas, methods and tips to produce fruit.

So, if I start with apples, it should come as no surprise. In Cornwall, I had an orchard with most of the apples all grafted by me and with lots of varieties, so I could literally go to the orchard from August until December and pick an apple off a tree. Some we ate, some we used for cooking, some we juiced for juice and some for cider. They are so versatile and if you do a little bit of research, they are really easy to grow. In France nearly all country houses have an apple tree or a little *verger* (orchard).

ORCHARDS, TOP FRUIT & SOFT FRUIT

Growing fruit is one of the most rewarding things you can do. There really is nothing like eating a home produced apple or peach straight from your own tree. In France the climate in the country as a whole means you will have no problems growing apples, pears, cherries, and most nuts anywhere in the country. South of the Loire vines, figs and peaches will produce good crops, and on coastal areas in Provence you may be able to produce oranges, lemons, almonds and pomegranates. Obviously, if you are located in an area of fierce winds or late spring frosts you may not be as productive, as the frost damages the flowers and pollinating insects do not fly in very cold weather, and you will probably have to make provision for rainwater harvesting to ensure your fruit have enough moisture during drought times.

Grow the same species of fruit trees together as this aids pollination, and allow space between trees to allow for ride-on mowers or wheelbarrows and the spread of the crowns of the trees themselves.

Apples can be one of the easiest fruits to grow requiring little effort and can produce a good crop of apples in just 5 years. The first thing you need to consider when choosing trees is how much land you can spare for an orchard and how large you wish trees to grow. I would suggest for a field system, traditional orchard (*verger*) to choose M25 rootstock, and for medium to large gardens a smaller MM106 or M26 rootstock. The rootstock (*porte-greffe*) choice governs both the height at maturity and the vigour of the trees. Larger trees take longer to mature and produce fruit, taking 10 years to reach maturity. Smaller rootstocks, such as MM106 will be fruiting in 3 years and almost mature at 6 years old. France tends to use M9 rootstocks which sadly are susceptible to fireblight.

For apples to grow as standards in a *potager* or for espaliers or cordons choose M26 or M9 rootstocks. Espaliers are small trees trained flat against a wall. Cordon trees are also trained to produce fruit from fruiting spurs along a single stem or trunk. These smaller growing trees will need support and training and will take longer to establish than the more vigorous rootstocks.

Choose varieties based on your taste, disease resistance and the season you would like to harvest. You will need at least two different varieties to ensure cross pollination. So, if you just wanted two trees, then perhaps two mid season, or one mid season and one slightly later will satisfy your needs. There are so many different varieties to choose from that I can't possibly list examples here, but it's a pleasant occupation for a winter's evening, working your way through plant catalogues and checking for the ideal apple.

Apples for eating; what we would call 'dessert' apples are called *pommes* 'à couteau'. There used to be over 4000 apple and pear varieties in France, and you can still find some of these old varieties by checking specialist *pépinières*, such as Pépinières Dumont in Aube; or Pèpinière Jouve-Racamond in Bouches-du-Rhone. Find out more about old apple varieties at **www.croqueurs-de-pommes.asso.fr**

Buy from a decent nursery (*pépinière*) or farm cooperative and plant as soon as possible in the year, or when you arrive except when it's very wet, frosty or very dry. Stake your tree properly and make a paper or computer file map listing the variety and when you planted it. To my shame, I did not do this in my first orchard and had to wait five years until they produced fruit for me to be able to identify individual trees. Having too many fingers in too many pies and a woolly memory on occasion is a pain.

If you have planted a one or two year old tree you should not have to prune or interfere with it, unless you need to water in the first summer (and really water it well every few days in dry spells in the evening). Pruning in late winter (I do mine on New Years Day) will allow you to remove any dead, diseased or rubbing twigs and encourages more growth. Summer pruning will encourage fruiting. That's all you have to do for the first couple of years. There is more information on pruning at the end of this chapter.

If you have enough pollinating insects and don't have fierce spring winds you should see your first fruit by year three. Don't allow too many fruitlets to form this young, it will only encourage the tree to produce small fruits and not to concentrate on growing and establishing itself.

The only insect pests I've seen on apples here in France have been moth caterpillars in the foliage. Codling moth and ermine moth can be an issue in fruit trees, creating webs and tightly rolled leaves. I simply take a bucket and clip off the affected webbed areas when I see them and then burn the leaves. If you have young trees you could spray with pyrethrum or permethrin if you desire. Grease bands (*bande de graisse*) placed around the trunks in autumn may prove a good deterrent. Keep grass short around apple (and other top fruit) tree trunks to discourage hibernation of pests.

Harvest as soon as the fruit is ripe. Early varieties do not store for long and are best eaten, juiced or sliced and frozen for pies. Later varieties seem to last and can be stored in flat trays wrapped in newspaper to separate them. Pomace from any juicing can be used to supplement animal feed rations or offered to other local pig producers in exchange for some sausages or pork when it comes to processing time. It not only contains a high sugar content, but also is full of protein.

PEARS

I have a wonderful memory of our huge, traditional perry pear tree in our garden in Cornwall. It was the first fruit tree to flower right after the hedges were covered in sloe blossom, and the smell of the pear blossom was heavenly. It also provided much needed early pollen and nectar for the honeybees and gave us a good crop of hard perry pears for making perry (a fermented alcoholic drink like cider, but made from pears).

Pears like apples have many varieties and have the same growing requirements and pruning needs as apples. Some varieties will produce too much fruit for the branches to bear and you may need to remove a lot of fruit to prevent the weight breaking the branches. Some varieties are also better trained on a wall as an espalier, with wiring supporting the fruit as they grow.

Traditional varieties such as *William bon Chretien, Doyenne d'hiver* and early pears such as *Saint Jean*. Every old village will have a Curate's pear tree (*poire-curé*) near the church, which at least gave the poor curate something to eat if his collection on a Sunday proved inadequate in hard times.

Again, prune to remove any dead, diseased or rubbing growth, and tidy to the shape you require in winter. Pears are hard to keep unless you freeze them, so usually one or two pear trees will produce abundant fruit for your needs.

PLUMS

Plums come in many shapes, sizes and colours. In July, little golden mirabelles start the season, followed by sweet cherry plums. These small tomato-like fruits are ideal for jams and pies, so collect and freeze in small freezer bags, if there are too many to eat. If you don't fancy them in pies, there is jam or even in a bottle with vodka or gin to produce a pleasant winter tipple.

As the mirabelles finish, the plum season really starts with blue and purple plums and greengages becoming ripe. Personal taste is really key here. I'm a fan of golden mirabelles for eating, and purple plums for crumbles. My husband just loves eating plums straight from the tree. One thing is for certain; I like to grow them away from the house as they attract wasps, and I'm not keen on these pests.

Plums, like cherries, peaches and nectarines are stone fruit and must be pruned (if required) in May to prevent the incidence of Silver Leaf in the tree. Just remove excess growth back to maintain a good shape and encourage the production of fruiting buds next spring.

PEACHES & NECTARINES

Having never been able to grow anything like this in the UK, I was delighted to see that our neighbours all have a nectarine or peach tree. Naturally, I went straight out and bought one of each. They have established quite well and in the second summer after planting we had peaches and nectarines from our own trees. I remove any blistered leaves in spring and spray with Bordeaux Mix against peach leaf curl. My neighbours hang little net bags of eggshells in the branches of their trees. Either way, we don't seem particularly susceptible. There really is nothing nicer than picking a ripe peach from your own tree.

CHERRIES

Cherries seem to produce really well in France and many potagers and gardens are bright with the pretty blossoming trees in Springtime. Birds can be a nuisance and beat you to the harvest, but tall trees are almost impossible to protect, even if you hang silver streamers in the branches. *Morello* and *Stella* are both self fertile varieties. Wild cherries also provide some fruit and make a pretty edition to woodlands if you are lucky enough to have one. Pruning if required is in June (like all stone fruit) to discourage Silver Leaf disease. Go easy on any pruning, removing dead or crossing branches where necessary.

EXOTICS (Pomegranates, Feijoa etc.)

If you intend to relocate south of the Loire valley divide or find yourself in a tropical micro-climate you may want to try some exotic fruits. Pomegranates (*grenade*) can indeed be grown in France, (I have two different varieties) and produce the most beautiful flowers. I am yet to collect any fruit from them as they are still young, but I have seen large trees in a nearby village laden with red fruit in late autumn.

Persimmon (*kaki*) also grow well here and the pretty orange fruits remain on the tree well into December when most other trees have lost their leaves. This gives the impression of a bare tree covered with orange Christmas baubles! These are tall trees though, so plant with care.

Feijoa is a relative newcomer to French gardens. It is a large evergreen shrub with green leaves with a grey or white downy reverse, and the most beautiful flowers of purple and white with long red stamens. It is hardy and can withstand winters of minus 15. As the plant is a native of Brazil, there are no hummingbirds in France to pollinate it, so you will have to get a small soft artist's paintbrush to physically transfer pollen from one flower to another or you will never have fruit. The fruits are oval, green and firm, and very perfumed. The taste of the fruit is something like a gritty guava, but they don't keep well, so try to infuse in vodka for a month or so to make an unusual and delicious liqueur.

Passionfruit (*grenadille or couzou*) is a large, climbing perennial vine and may or may not provide you with much fruit. It seems that the flowers appear to provide enough enjoyment for gardeners, although you can obtain some fruit in the southern areas of France. Fruit usually appears from August to October in the third year of growth and will drop when ripe. To encourage more flowering and fruiting, prune excess growth in early spring and add a rich mulch to the roots. They like sandy, warm soil and dislike cold winds and frosts. They are susceptible to fusarium wilt so buy grafted varieties rather than grow from seeds. They have a short life of between 5 and 7 years.

Kiwifruit (Actinidia deliciosa) have proved to be a great success commercially in South West France and France is now the 5th largest supplier of kiwi fruit in the world. They require warm, neutral soil and need a strong support to climb over as the vines are large (up to 5m long) and heavy. They can survive up to -15 in winter, and harvest is from October to end of November. Fruits can be cold stored for a short time. You will need a male and a female plant, although the variety *Jenny* is supposedly self fertile. A French variety *Montcap* and the variety *Hayward* are available to buy as either male or female plants.

Olives, if you can ripen them can be harvested and taken to a nearby mill to be exchanged for oil if you're area produces and processes them. Or you can collect and brine them for eating as aperitifs. The blackbirds strip our trees just as they ripen so I haven't managed to do this yet.

SOFT FRUIT

There is such a multitude of soft fruits that can easily be cultivated and found in France that I feel it is pointless to really examine them in detail. Personally, I grow what is expensive and hard to find in the markets or supermarkets. Alpine strawberries can be found nestling under some taller plants in my raised borders where they appear to be quite happy and pest free. If I want strawberries out of season, I may grow them to crop early in polytunnels and then some later ones in a cold frame. It's harder to force cane fruits such as raspberries to do this, but if you want a succession of fruit then look out for early and later fruiting varieties to extend the season. The potager is handily situated next to the shed with water butts which give added moisture during the hot summers.

Currants and raspberries are planted at the back of my potager in rows, where they shade each other a little, and are close to the water butts next to our shed, but I have to say my blackcurrants aren't really thriving here in hot south Dordogne. Raspberries are self fertile and good varieties include *Glen Ample, Autumn Bliss* and *Allgold* (a golden variety). Gooseberries are either cooking or dessert varieties. *Careless* is a good cooker, whilst *Leveller* or *Martlet* (a red fruit) are good dessert varieties.

Try and protect all soft fruit from predators. I like tall bamboo canes with plastic cups on top covered with a lightweight bird-proof mesh. Slugs and snails can be deterred with crushed egg shell, coffee grounds or trapped in upside down empty orange skin halves. I remember in Cornwall searching the fruit beds at night with a torch and bucket with some salt in the bottom and picking them by hand.

Prune fruited raspberry canes to the ground after fruiting and leave new young growth. Blackcurrants need removal of old central wood in winter to allow new growth. Gooseberries need any tangled wood removed, again in winter to allow air and light.

MELONS

There are a few varieties of melon grown in the warmer regions of France, either outdoors or in polytunnels or cloches. The most famous and popular is the cantaloupe charentais. This is the medium size round melon with rough lacy skin and a tender orange flesh. There are other local varieties available and they are relatively easy to grow, but demand a little space and a little more watering until they have stopped growing. If you are in a cooler region, try them in a polytunnel with top ventilation and water well, avoiding the leaves.

When you are sure all danger of frost has gone start to plant outdoors. Make a little mound of well-rotted compost on top of the soil in a sunny well fertilised plot. Plant 3 seeds per mound about an inch deep. Space the mounds just over a metre apart. Water and then cover the mounds with a few inches of straw. As the vines grow keep the soil moist (but avoid wetting the leaves). Cut back the main leader after it has thrown out a couple of laterals and restrict flower production to three female flowers at a time. Remember you also need a couple of male flowers to pollinate. The melons will grow and are ripe from June to September. Check ripeness when the tendril nearest the fruit turns brown. Also, the skin will have hardened with its laced appearance and the stem will have 'split' away from the globe slightly. The melon should feel heavy. Ripe melons will keep for 2-3 days if hung in a net in a very cool room.

FIGS

We have inherited two mature fig bushes, which I prune back by about half every summer. The first gets sun all day and figs are ready to harvest in mid August, the second has sun until around 3pm, but is ready to harvest first. You'll know when they are ripe when the fruit 'droops' slightly and 'gives' when gently pressed. We have no idea of the varieties, but have to say they are delicious. Foot long cuttings are easily taken in spring and simply shoved into the soil about half their length. In autumn they have formed roots and are ready to pot up and gift to friends. My recipe for fig chutney is at the end of this chapter.

WALNUTS & OTHER NUTS

The walnut *(noix)* was originally native to Persia and brought to France by the Romans. The climate all over France is suitable for their cultivation, but commercially the Perigord seems to be the favourite growing area. The fruit can be eaten fresh or dried, pickled and to produce oil, and the wood is highly sought after by cabinetmakers and as firewood.

Grafted named varieties can start producing nuts after 3 years, compared to 15-20 years for a self sown walnut. They are large trees and unless you have a fancy to own a Walnut grove, one tree should provide sufficiently for your needs. *Franquette* is the traditional variety with a moderate yield and good blight resistance. *Fernor* is a modern cultivar with similar blight resistance and a good tasty crop.

Harvest time is October and November when ripe nuts fall from the trees or are shaken by a tractor in commercial walnut groves. Nuts are dried in the sun in large flat trays. The nuts are then shelled, usually by hand, resulting in brown stained hands. Old villagers will reminisce about the *veillées* (evening gatherings) in the old days, where everyone used to gossip and pass the time whilst shelling nuts. There is a good article on the recent commercial success of nut production in France here Nut Production France **http://irrigazette.com/ fr/node/1018**

If you are looking for a specialist nursery to purchase grafted nut trees look here: **www.pepinoix.com/en/walnuts-nursery/**

BEWARE: THE OUTER GREEN HUSK TURNS BLACK AND SLIMY ON THE GROUND. IT IS HIGHLY TOXIC TO ANIMALS AND EVEN A FEW CAN PROVE FATAL IF EATEN.

Commercial hazel nut (*noisette*) production in France is centred in Aquitaine and the Midi-Pyrenees areas of France, but hazel will grow well in the moderate, cooler and moister areas of France and are happiest in cool soils at the edge of woodlands and in hedges. They make an excellent windbreak or screening for an orchard. The shrubs do grow fairly large, but are frequently coppiced to maintain vigour and to harvest the pliable timber they produce.

There are two species, Corylus avellana (hazel nuts) and Corylus maxima (Filberts). The two species produce slightly different shaped nuts and have differing growth forms. Corylus avellana '*Cosford*' is a good, self fertile variety with good disease resistance. '*Fertile de Coutard*' is a French variety with large nuts. '*Pawetet*' is another high yielding variety. Both these last two varieties are not self-fertile and require a second variety as a pollinator. Of course, there is nothing to stop you planting English varieties such as *Kentish Cob*, *Cosford cob* or the purple-leaved filbert. Propagation is best from layering. The main pests are squirrels and nut weevils.

You can of course forage for hazelnuts in local woods or roadsides, and perhaps plant wild individuals either in a hedge or a stand alone bush. With beautiful yellow catkins from late winter to early spring, they remind us that spring is around the corner and summer not far behind. By August, some immature nuts begin to fall from the trees. In many parts of France these are considered a delicate *aperitif.* All that is required is to peel back the jacket, crack the not-so-hard shell and eat the cream coloured kernel. By September, the nuts have ripened and fall like rain from the bushes for almost a month. Pick only when ripe as the immature nuts will shrivel. Dry in the sun or near a wood burner, then crack and toast to remove the bitter skins. These are excellent with a tot of port or a liqueur as part of a little supper of cheese, bread and grapes.

PRUNING

Firstly, we have to divide the species of trees to be pruned into distinct groups. Apples & pears in one group and stone fruit such as plums, cherries, peaches etc in the other. Yes, you may have to prune other species such as nuts or kiwi, but basically prune these others in the dormant period to reduce congestion and weight load on branches then just remove dead, diseased or prune to shape.

Apples and pears are mainly pruned in winter when the leaves have fallen off and the tree is dormant. The key points to remember are that pruning is only necessary to remove dead, diseased or rubbing branches or to top very tall trees or to open up a congested older tree. When pruning I leave everything I've removed at the bottom of the tree to remind me that there is always next year. Every cut gives the tree a little shock, so less is more, so to speak. You *can* summer prune if there is an abundance of growth and no fruit formation.

As apples are my personal favourite fruit and one of the easiest to grow and enjoy I am sharing below exactly how to make juice and cider. For either or both all you need are fresh apples, some clean bottles and a way to pasteurise. Now, pasteurisers are expensive, so even a large deep pan can be used to bring water to the correct pasteurising temperature, which you can measure with a thermometer.

Apple Juice and Cider

Re-using clean glass crown cap bottles (such as empty beer bottles) in 300 or 500ml size is a good size for both juice and cider. You can also buy 3L or 5L bag in box wine or cider bags for larger quantities, but will need a bath pasteuriser to pasteurise these. A tall thermometer is useful to track the temperature and a clock alarm to tell you when time is up.

Juice and Cider can be made from almost any type of apple. However, you will get a more complex flavour if you use a mix of apple types – dessert, cookers and cider apples. Apples can be divided into having four tastes - Sweets, Sharps, Bittersweets and Bittersharp. The last two are found in cider apples because of the high concentration of tannin. If you don't have access to cider apples, you could try adding some crab apples to supply the tannins, but it's not essential.

Apples are ripe when the pips are brown and when the skin 'gives' a little when pressed. For juice you must use unblemished apples, preferably picked from the tree and not taken from the ground; especially if there are animals that have access to the land. Cider is more forgiving and commercial cider makers readily collect windfalls.

If you are making juice you don't need a hydrometer, just wash the apples prior to scratting (pulping) and pressing. As you press the juice will exude and quickly turn a dark golden colour. If you wish to preserve the pale yellow 'fresh press' colour you must add Acetic Acid (5g per 10L of juice) powder whilst you are scratting the apples and mix well in immediately.

The amount of juice you can get depends on the apples, how dry a summer you have had and what sort of press you use. We use a rack and cloth press with a 10 tonne bottle jack. There is no need to add sugar or water to the juice. Juicing equipment is expensive! Ensure you dismantle both the scratter and the press and thoroughly clean both and dry before storing. This often takes as much time as pressing the fruit itself.

Once you have pressed your juice into a clean container, cover with a lid and leave overnight. The pulp will settle to the bottom and you can then use a siphon to fill individual bottles to the bottom of the neck of the glass bottle. Cap the bottles with a crown cap and place in your pasteuriser (or large deep pan full of water) and heat until the temperature reaches 70 degrees. Put the timer on for 20 minutes, keep the temperature steady, and once the 20 minutes is up remove the bottles carefully and lay on a protected flat surface (a towel on the floor is sufficient) to allow the very hot juice to fill the neck and underside of the cap and pasteurise this also. The bottle will cool and can then be stored. Juice correctly pasteurised is good for 18 month to 2 years.

To make cider the process is similar. Juice the apples and collect the juice in large containers. Do NOT add Acetic Acid. Once juiced, measure the amount of sugar in the juice to ascertain how alcoholic the finished cider will be. To do this you need a hydrometer. Normal starting readings are between 1.045 and 1.07 (which will give a finished ABV of around 6% and 8.5%). If the initial specific gravity (S.G.) level is less than 1.045 (perhaps you've picked your apples before they were ripe), then you need to add some sugar to bring it to 1.045. Add sugar in increments of 15g, mix well and retest.

So, you have the juice and it's housed in a suitable clean vessel for fermentation. Demi-johns are ideal for beginners or small quantities as they can be filled to just above the shoulder and the narrow neck is ideal for a rubber bung fitted with a modern airlock. You can either allow a wild yeast fermentation to take place or you can add a sachet of cider yeast (not bread yeast please). OR - You could also add some Campden tablets at this stage, leave for 48 hours and then add your chosen cider yeast, but we are getting complicated for the scope of this book.

A plug of cotton wool for the initial, sometimes fierce fermentation can be removed and replaced with a proper airlock after a fortnight or so. This allows carbon dioxide to escape but prevents oxygen entering. Oxygen will turn your cider to vinegar. Fermentation will start within 48 hours if you have added yeast, slightly slower for wild yeasts to take hold. Bubbles and sometimes froth are numerous at the start. As fermentation progresses the rise of little CO_2 bubbles in the airlock slows right down. When you think it has almost stopped bubbling altogether, try siphoning off a little and testing with the hydrometer. If it reads higher than 1.005 then it has become stuck and you may need to add either a nutrient or some yeast or both. If it reads 1.005, then you are ready to rack off into a new sterile container.

Try to do this as gently as possible and leave as much yeast and sediment as possible in the original container. Put the siphon into the new container to avoid aeration as much as you can. You will see that the level of cider in the new container is much less now. Do not top this up with apple juice or you will simply start a new fermentation. You can top up with water to a level as near the top of the neck of the container as possible. You no longer have juice remember, and cider and oxygen are mortal enemies. Replace the bung and airlock.

You may, as a result of moving the liquid around get a little more bubbles and fermentation, but this will settle quickly. You can now store the finished cider for maturation or bottle and pasteurise it. Leave to mature for at least three months, although it will taste much smoother after six months.

If you bottle, then leave some airspace for expansion, cap with a crown cap and pasteurise in a bath pasteuriser or large pan for 20 minutes at 70 degrees. Carefully remove the bottles and lay flat on a protected surface (I use an old towel) to allow the hot cider to pasteurise inside the neck and cap of the bottle. When cool can be stored for 18 months to two years.

This is a very simple explanation, and you may have questions or problems. I will be writing a book on apples and cider making soon, but there is a lot of information readily available on the internet or you can visit my Facebook page: **Sustainable Smallholding** for more information.

SOIL, COMPOST & WATERING

One of your first tasks when looking at producing crops or even keeping livestock will be to get to know your soil. Is it thin and poor or rich and fertile? What pH is it? Has it been cultivated recently? All life on this planet owes itself to the presence of 6 inches of topsoil and the fact that it rains, so don't underestimate the importance of the soil.

Buy a small pH test kit (*testeur d'acidité*) and take some samples from different areas of your land. In Cornwall we had thin, stony soil with a pH of 6 in the land near the access road and deep rich loamy soil with a pH of 6.5 in the rest of the land. We had planned to plant our orchard at the front near the gate but the soil was too poor and the depth only 9 inches before we hit bedrock, so instead we planted it in the rear paddock. You are looking for a pH of around 6.5, but you can improve soils with pH from 5 to 7. With 7 being pH neutral, 5 indicates acidic soil and anything above 7 indicates alkaline soil.

Some of the plants growing on the land can also give you clues about your soil. For example, hypericum (St John's Wort), periwinkle and scarlet Pimpernel indicate a dry, sandy soil which may be on the acidic side. Gorse, heather and spruce indicate a peaty, damp acidic soil. Yellow iris, goat willow, cuckoo flower, and meadow sweet are likely to be seen on marshy, clay soils with a neutral pH. Hazel, elder and horsetails tend to favour cooler chalky, alkaline soil, whilst dry alkaline, stony or chalky soils will have poppies, lavender, herbs, wild clematis, and hawthorn growing happily. Where you see nettles and burdock you will find soil rich in nitrates.

You can improve drainage and fertility by adding in gravel or flint and by adding organic matter such as animal manure, composted straw or grass and this will the soil structure and also the pH. Applying lime to barren fields will reduce the acidity, but you may have to apply annually for a few years to get the pH level closer to neutral. Rome wasn't built in a day, so remember not to expect instant results.

I have to say I am now a supporter of No-Dig cultivation. This method is championed by many growers worldwide and I can thoroughly recommend having a look at the website of Charles Dowding. Charles has written many books and articles on the subject. You can see his site here: **www. charlesdowding.co.uk**

The method basically consists of piling a compost layer on top of your chosen area to be cultivated. This then smothers weed growth and at the same time feeds the soil. In a few months you can plant straight into the area. Topping up the compost keeps the soil fertile and weed free. If you start with a six inch/25cm layer on top of uncultivated or new soil in November, and then top up in January, you should be ready to plant in the spring. You may have to remove the occasional persistent offender, but by and large this method works. This method means more cropping, less weeds and definitely less effort. The compost can be a mix of thick cardboard, old animal manure, straw, leaves and grass clippings.

Of course this method won't be suitable for field size plots, but for vegetable gardens and *potagers* then it's certainly effective. For larger vegetable plots initial rotovating may be more suitable, and for fields then you really are talking about having someone to come and plough and harrow before planting.

Compost can easily be made from spent animal bedding, animal manure, grass cuttings, vegetable waste, cardboard and leaves. Avoid bones, meat or oils from the kitchen or you will attract rats. A heap in a shady area at the bottom of the garden will start to rot down and provide a rich brown compost for use in the vegetable garden, potager or to mulch flower beds. In fact two heaps next to each other allows you to fill one and then periodically toss this into the second heap, thus aerating the matter and putting new material to the bottom of the heap. Compost should be ready within 6-9 months.

While we are still on the compost theme, let's talk about fertilizer. Now, I don't like the idea of buying in sacks of fertilizer and frankly I'd rather produce my own, and so can you. The most important elements that plants require are Nitrogen, Phosphorus and Potassium, shown on a lot of ready made brands as NPK. Now did you know that human urine produces *exactly* these elements in *exactly* the correct proportions that plants require? So, if you fancy having a lidded bucket in your loo especially for this purpose you can then take that bucket and dilute your pee with 10 parts water and use this to fertilize your fruit, vegetables and borders. Empty the bucket daily and swill out with clean water or your bathroom or loo will smell.

If you don't fancy this idea much; and my husband thinks we are weird enough, you can make a good liquid fertiliser by using animal manure (sheep, chicken, or horse is great) in a Hessian sack suspended in a water butt. If you don't have access to animal manure then stuffing a water butt with comfrey or nettles will produce another lovely fertiliser. This does have a pungent aroma, it has to be said and you must keep the butt topped up with water and covered (which also stops mosquitos breeding in the water). Dilute until it looks like weak black tea before using.

As the effects of climate change are all around us, we as gardeners and producers need to consider harvesting and using rainwater to water our fruit, vegetable gardens and potagers. Traditional water butts (*récupérateurs*) are available to buy from farm stores, building depots and garden centres, and can be fitted to collect water from your down pipes simply and effectively. Sizes vary from 150 litres capacity to 300 litres. A lighter colour will absorb less heat in summer and be less prone to buckling in shape slightly.

The roofs of all farm buildings and sheds should be used to collect water. Even a small roof space can collect a huge amount of water. Our tractor shed in the garden has gutters and downpipes leading to two 1000 Litre IBC containers, which don't look as pretty, but they collect a lot of water and are situated right next to the *potager*. Site any *récupérateurs* or water butts about a foot off the ground on solid concrete blocks, so you can place a bucket or watering can under the tap easily. We have changed the tap at the bottom to one that can connect to a hose and uses the weight of the water and gravity to force water through the hosepipe.

POTAGERS

A *potager* is an ornamental French kitchen garden, where traditionally fruit, vegetables, herbs and flowers all intermingle informally. Today potagers tend to be romantic and the French version of English cottage gardens. This informality also means that weeds will tend to thrive, but remember a weed is simply a plant in the wrong place: dandelions, for instance make great beer and wine and the leaves are great in salad.

Planning a *potager* needs an evening of contemplation and paper and pencils. You need to have a look at the site itself. Is it sheltered or windy? Is it near the kitchen? How big do you plan it to be? What shape? There are lots of things to consider. Firstly, if the *potager* is the only place you will grow herbs or vegetables then it needs to be bigger than a cartwheel. Smaller is best if you intend to be away from home frequently, or perhaps struggle physically with digging or lifting or just don't fancy much labour then smaller is best. A sunny, sheltered spot always produces more, and having a handy water supply nearby will save time, labour and frustration in dry spells.

How are you going to cultivate it? If by hand, can you manage the labour required or will you adopt a 'no dig' approach mentioned above? If you are going to rotovate or even use a small tractor then you need to consider access issues.

Raised beds can help if you have wet soil or if you are physically challenged. I like the height of 2 foot off the ground. It's a good height for me to sit in a chair and weed or if I have children helping me, and by the time I'm elderly,

I may need a wheelchair, who knows? Broad paths are a great idea at the start of any design project. 3-4 ft wide level paths allow you to manoeuvre a wheelbarrow or wheelchair in and out. A gravel or mulched surface makes for a good permanent path.

Remember, a traditional potager is a mixed fruit/vegetable/flower/herb area, so including some height can make a great visual impact and increase your productivity. Small fruit trees, such as apples, plums, cherries, peaches or nectarines make a great addition. Climbing roses, vines and tripods with climbing beans and nasturtiums add to the picture, and free standing tall herbs such as fennel can also find a place.

Then choosing a shape may be important if the *potager* is to be seen from the house. Tall plants in the centre or towards the back, with smaller plants at the edges and front will allow you to see everything from the front and allow filling of bare spots when you have harvested anything. Don't make the depth too deep unless you plan paths to allow you access to your produce at harvest time. Geometric shapes were the traditional choice, but I've seen spirals with gravel and also checkerboards and triangles used well. You could even use your *potager* as a hedge to divide parts of your garden off.

The choice of flowers, herbs and fruit and vegetables is up to you and may depend on whereabouts in the country you live. It makes sense to try and grow items you find expensive in the shops, or that you use a lot. Personally, fresh salad is a must, and I grow and use edible flowers too! I grow lots of onions, beetroot and haricot beans, *cavolo nero* and pumpkins for winter soup. Unfortunately, my area is too warm really for blackcurrants, so I have to buy blackcurrant jam at the supermarket until I find a local supplier.

There are many examples of *potager* design and cultivation on the internet. One great site is a permaculture farm, called

Le Bec Hellouin in Normandy: **http://fermedebec.com**

If you wish to have normal vegetable plots or beds, of course this is your choice. We have tomato plants dotted around our raised flower borders so that if we have guests with children in our *gites* they can go collect tomatoes within easy reach.

What to grow? Well, my plans had to take account of the climate and temperature here. Sweetcorn gets routinely attacked by insects and so I don't grow it now, and I can grow melons here if I have enough reclaimed water in the summer as we are often subject to severe drought. My main factors are:

Grow crops that we both enjoy
Grow crops that like the soil & climate
Easy and disease resistant
Grow things that are expensive in shops

HERBS

Herbs are widely grown and used by the French in cooking and around the home. In the countryside and villages, lavender is cheek by jowl with fennel, basil, chives and other herbs. Most enjoy a Mediterranean climate, which means dry or gritty soil. The reason so much lavender fails in the UK is because the soil is too wet in winter, so if you have a cold, heavy, wet soil, add a handful of grit in the planting hole to add the drainage of water.

Most herbs will also self sow if left to go to seed, and mature plants are short lived. Cuttings from sage, basil and rosemary are easily propagated in early summer: just break off a piece and stick it into the ground. It should have rooted by autumn and can be potted up to plant elsewhere. It is well known that many herbs repel creatures, for example, lavender in the henhouse will deter lice and mites, and rabbits hate rosemary.

Beware of allowing invasive herbs, such as mint, to take over. Unless confined to a pot, they will quickly infest any beds. Too many fennel seedlings can simply be pulled up when young and used to flavour food such as fish.

Herbs can be dried and crumbled into clean jars or frozen into ice cubes for use in cooking. Lavender cut and dried upside down in bunches repels flies from the house and can be used for so many crafted items such as pillows to aid sleeping, in a wheat bag to heat in the microwave to ease muscle pain or in small bags hung in wardrobes and in drawers to deter clothes moths.

POLYTUNNELS & GREENHOUSES

Greenhouses (*serre*) are quite rare in rural France, and their use is confined to extending the growing season or starting baby plants off. In the summer they will require heavy shading and lots of ventilation. In autumn they are useful to hang and dry a variety of onions and herbs. If you intend to use them to over winter delicate plants such as oranges or other pot plants you will require a source of heat.

Small poly-tunnels are, on the other hand quite common, and have a multitude of uses, including lawnmower storage, or for hanging out laundry if fitted with a central line. Remember that you will need planning permission for any greenhouse or poly-tunnel with a concrete base, or for tunnels over 1.80m high (*déclaration préalable de travaux*). Check at your local *mairie*'s office for regulations in your area.

They are very useful for propagating and hardening off young plants and vegetables before the frosts have finally disappeared (traditionally around the 10th-12th of May), but they are also used to grow early and late crops of potatoes, beans or peas, salad crops and soft fruit such as strawberries.

Remember the soil will soon tire and need generous applications of compost or manure to keep fertile. You will also need to practise some crop rotation to avoid problems and be vigilant in your pest control (both insect and rodent). Also, the fact that no natural rain can enter the poly-tunnel means you'll need a source of water nearby; either from mains, a spring or stream, or a water butt or *récupérateur*.

CROP ROTATION

French gardeners and farmers all recognise the importance of crop rotation to prevent a build up of pests, diseases and to avoid a deficiency in certain nutrients. The plot or poly-tunnel is divided into 3 sections; the first containing greedy feeders (aubergine, peppers, tomatoes, cucumber, melon and roots, the second one for less greedy feeders (brassicas, leeks, garlic, salad crops, sweet corn) and the third for poor eaters and leguminous plants(peas, beans, carrots & parsnips). Before planting the first group, ensure you have spread and dug in a good layer of well rotted manure or compost. Repeat this at the end of the third season. Then simply rotate the beds.

In season one, bed one has the greedy feeders,
in season two bed two has the greedy feeders (and you've fertilized before planting)
in season three bed three has the greedy feeders (again, fertilized before planting).

PESTS, DISEASES & CHEMICAL CONTROL

Whilst you can't do much against certain pests such as rabbits, deer and wild boar except protect by fencing, smaller pests such as rodents can be kept at bay with the handy addition of a cat, and by ensuring that you keep sheds and outbuildings clean and free from spilt grain or animal feed (the best place to keep these in secure metal dustbins with tight fitting lids).

For your fruit and vegetables the main pests are aphids (*aphides*), caterpillars (*chenille*) and slugs (*limace*) and snails (*escargot*).

Insecticides can be effective but have adverse effects on the soil and are a danger to other wildlife and pets. Nematodes are effective, especially in greenhouses and poly-tunnels and are biologically acceptable as you are using nature to fight nature. Planting certain plants next to others as companion planting may protect crops from infestation. For instance, marigolds planted between tomatoes and potatoes deter whitefly. However, be careful when trying companion planting – some of the suggestions are just daft. For instance, it is said that planting peppermint beside cabbages will deter butterflies and therefore caterpillars, but who wants their vegetable plot overrun with very invasive mint? A natural insecticide can be made by steeping chopped up young nettles in a bucket full of rainwater and leave to ferment for a week or so, then dilute and spray to deal with whitefly.

And age-old methods are effective - such as scattering soot between rows of young carrot to repel carrot fly or slugs, eggshells crushed and scattered around tender plants to deter slugs may work. Spraying diluted soapy water onto plants infested with aphids removes their protective layer and they will desiccate and die. Re-use empty window cleaning sprayers well washed out for this.

Whilst annual weeds can be hoed out and left to die in the sun, perennial weeds and tree stumps will always pose problems for the gardener or farmer. Whilst good old brute force is an excellent method of rooting out brambles and the like, facing the slog of dealing with this over a large area is a thankless task. Many of us would gladly turn our backs on this or turn to the guaranteed results of a chemical weedkiller.

There are two main kinds. Firstly, a contact herbicide, of which acetic acid (vinegar) or pelargonic acid are the main ingredient, kills the growth above ground but not the roots. These are fine for large areas of annual weeds. The second kind are systemic weedkillers. When you spray the foliage, the plant absorbs the chemical and draws it into the plant which then dies.

We are or should be aware that worldwide the use of chemical weedkiller (*désherbant*) is under scrutiny with one particular brand being singled out. France has banned all non-professional use of glyphosate-based weedkiller since January 2019. Professionals can still use it until 2021. A glyphosate weedkiller is systemic. However, there are alternative weedkillers available in France for dealing with perennial or woody weeds. These include Triclopyr as their active ingredient. So, if you need to tackle brambles, dock, nettles, ivy or tree stumps have a look for Vitax SBK brush wood killer, or products containing Garlon, such as Garlon herbicide, Bross Tueur Brushkiller. So far, these products have not been banned for non-professional use, but do please be wary of the use of chemicals especially if you are applying them close to areas which will be used for food production or in the vicinity of people or animals.

Before we leave the subject of vegetables totally, I need to explain the French belief of the system of planting 'by the moon'. We know that the moon influences tides and it is believed in France that certain plants must be planted at certain phases of the moon. Root vegetables and plants that require to grow slowly and not go to seed quickly are planted when the moon is waning (getting smaller); and fruit trees, shrubs and flowers are best planted when the moon is waxing (getting bigger). If you aren't convinced of this (and I have to admit to initially being a bit of a sceptic), then think of the amount of heavy rain falls immediately at or after a full moon.

Delphine's Chutney de figues

1.8kg green or purple figs
700g sugar
4 sweet onions
Handful of raisins
40cl red wine vinegar
1 tbsp fresh grated or fine chopped ginger
½ tsp grated nutmeg
½ tsp black peppercorns
1 tsp salt
½ cinnamon stick

Peel and chop the onions. Wash & cut the figs into quarters. Put figs, raisins and onions into a big pot or casserole with the wine vinegar. Bring to the boil and cook gently for 20 minutes. Add the salt, pepper, ginger, cinnamon, nutmeg & sugar and stir well until the sugar dissolves. Thicken over a low heat, stirring often - this takes about 40 mins. You can jar this in warm jars (inverting the lidded jar to sterilise the lid) or put into small containers to freeze.

Chapter 6
Machinery & Tools

Moving to rural France means you will need at least a toolbox with some of the normal type of tools, such as hammers, screwdrivers, spanners, pliers etc, but if you intend to do any renovation inside a building, fencing agricultural or horticultural practices you will soon need a much bigger set of tools, machinery and PPE (Personal Protective Equipment).

When we moved to France from our one acre smallholding in Cornwall, half of the huge lorry we hired to move our belongings was filled with my husband's machinery and tools. And I have to tell you we had sold off lots of machinery before we left! He rightly foresaw that we would need at least one welder, small hand power tools, a decent chainsaw and PPE, a push along lawn mower, a strimmer and a small trailer. A two-stage aluminium ladder has been brought into use numerous times. It can be stored hanging from the beams of your shed or along the wall for storage. A step-up platform is also useful.

Obviously what you will need will depend on what you intend to do, but if you already have these things then bring them with you. France is an expensive country and from hindsight I would advise you to buy quality. Yes, you may be able to borrow some items from a friendly neighbour or rent from a company such as Loxam. Remember you will have to pay a hire charge, delivery, and a damage deposit. If you hire any equipment, ensure you have checked it for damage or malfunction before you accept delivery; that you know how to operate and or refuel it and what type of fuel it requires.

If you have a wood burner (and most rural French homes have at least one wood burner) and intend to grow and cut at least some of your own timber as fuel you will need a decent chainsaw (*tronçonneuse*) and PPE (personal protection equipment), and ensure you wear it. Quite rightly, people are afraid of chainsaws – the damage they can do in untrained hands can be fatal. If you feel happier having a contractor to come and do this work for you instead, ensure they (and you) have sufficient insurance before agreeing to a price. You can check this via your bank, using the *chèque emploi service.*

If you decide to do it yourself, remember safety is paramount and ensure you purchase and wear Kevlar trousers or dungarees, stout steel toecap boots and a helmet with a visor. Ear defenders and Kevlar gloves complete the outfit. Get some training in using and servicing your machine. You can easily do this in the UK via agricultural colleges, Lantra or smallholder training groups such as DASH in Devon.

NEVER ATTEMPT TO USE A CHAINSAW TO CUT ANYTHING OVER SHOULDER HEIGHT

NEVER ATTEMPT TO REMOVE A WHOLE MATURE TREE UNLESS YOU HAVE ASSESSED THE POSSIBLE TRAJECTORY AND POSSIBLE KICKBACK.

Before you start your chainsaw at the start of the season, check that you have new and sufficient oil in it, replace old fuel, and check the chain for damage and sharpness. Tension it correctly and check the tension whenever you stop for a few minutes. When you have to refill with petrol, again check the oil level and that the chain is taut.

When the chain starts to become blunt or sticks, you can return to the workshop and replace with a new chain. Keep the used one, check it over and decide if it warrants being professionally sharpened. There are marks on the chain indicating the wear level. A new chain will cost roughly the same as having one sharpened (if you search the internet), so instead of using a chain weakened by excess sharpening, just buy a new one. We always buy two at a time to always have a spare and this prevents John being tempted to sharpen a well used chain.

A small trailer for use with your vehicle or sit-on mower will save you time and effort. You can drive your equipment up to the place in the woods you are working and at the end of the day, the logs can be placed straight into the trailer and taken back to stack. I always ensure I'm within walking distance to wherever John is working, and if he is working at height, I'm right there on the ground nearby. We take safety very seriously. We burn 2 tonne of wood annually; I'd hate to have to move that by hand. It is worth mentioning here the cost of buying in wood already cut to 50cm lengths. At a rough cost of €65 per stérè, and average use of 10 stérés for a winter in two fires, this soon mounts up. The ride-on pulls the trailer containing the logs to our drying out stack and the next day we will spend the morning stacking, which as the saying goes is 'the first heating from the logs'.

Trailers (*remorque*) are really essential for any sort of rural life in France. We have a small livestock trailer for collecting sheep, which is also useful for taking items to the *déchetterie* (the tip), for collecting large furniture items etc. A large braked trailer over 750kg will require both a *carte gris* (registration document) and insurance, if you ever take it on the road.

If you have lawn over half an acre, invest in a ride-on mower (*tondeuse auto-portée*). If you have over 3 acres of grass you really need to consider how to manage cutting this. A ride-on mower isn't suitable for this level of work, and you may have to consider a small tractor if you also do considerable agricultural work, but most smallholders don't need this investment. If a local farmer can come and top the pasture for you, then great. Even better would be to have it cut, dried and baled as hay. This will only be possible if the gateway or gates are wide enough to allow for the machinery and the field big enough to negotiate and make it worth the bother.

If you are considering buying a used ride-on mower there are a couple of things to look out for: TRY all the gears to ensure they are working and smooth, if it is a hydrostatic (i.e. automatic) model make sure it drives forwards and backwards. Operate the deck, blades and then check the condition of the deck underneath and all the drive belts. The machine should operate as expected and most modern machines have a failsafe seat cut-off. If the engine does NOT cut out when you leave the seat this may indicate that the microswitch has been bypassed. This is an important safety feature, so be aware.

We have cut one acre with a push along mower (*tondeuse*) and this took many hours; a ride-on can do an acre in just over an hour. An even better idea is getting a mulching blade fitted to your mower – this way there are no unsightly lines of mown grass nor collecting or disposing of grass clippings. Remember at certain times of the year you will be doing this every 4 days. Please use ride-on mowers with care on steep gradients, and try to cut vertically rather than mowing on the slope horizontally. The likelihood of rolling the machine and the severity of injury and death are considerably reduced.

A heavy duty, well balanced petrol strimmer (*débroussailleuse*) is another good investment. If you have a decent make that you bought in the UK, bring it with you, along with your push-along mower. Consider buying a harness for your strimmer as prolonged use is heavy on the arms and shoulders. Also consider getting strimmer attachments such as a polesaw (if you have an orchard, overhanging trees or tall hedges), or a hedgetrimmer attachment (for tall hedges). If you are buying a strimmer, it is worth getting one with a 4-stroke motor rather than a 2-stroke one as they are quieter, less prone to running problems and cheaper to refuel.

Do you have hedges? A petrol hedgecutter (*taille-haie*) is also on your list. (You can see this list is full of expensive boys' toys). Choose as light a model as you can with a short bar; it will be much easier to work with and you will be able to work for longer. If you are going to do cultivation in a vegetable garden or plant an orchard, consider using a rotary tiller or rotavator (*motobineuse* or *rotofraise*). A self-drive model with wheels will turnover soil faster than a human can, although you may have to go over fresh uncultivated ground a few times to get the blades into the soil. The machine is basically an engine attached to a drive shaft with a set of rotating digging tines or blades. For best results only consider models with a minimum of 5hp engine.

WELDERS

There are a few types of welders; however, as a smallholder you are likely to be looking at an arc welder which also known as a stick welder. Two main types would be a workshop based oil filled welder, which runs off mains electricity; or a fan cooled inverter, which can be portable and run off a generator or used in the workshop off the mains electricity.

With a welder you can basically join metal together with intense heat. They are useful for repairs to gates, fencing, machinery and trailer repairs. To do most smallholding repairs you will need a welder capable of 180 amp output, and suitable rods. Rods must be kept in the container they arrive in and kept dry and free from moisture.

You must not try to repair lifting equipment or trailer tow hitches unless you have adequate certification in welding for these jobs. Nearly all metals can be welded, with the development of specific welding rods. Do not, however, attempt to weld previously galvanised metal. This galvanisation must be completely removed 25mm either side of a break prior to welding as it is extremely toxic as a gas and can result in hospitalisation. It can be recoated with a cold galvanise spray afterwards.

Always clear your working area and inspect it prior to welding. Look out especially for fuel containers, oily rags, varnish or flammables and debris. If welding, for example your ride-on mower, ensure you disconnect the battery beforehand to prevent damage. Vehicle fuel tanks and pressure vessels are best left to the experts.

To weld safely, you must have adequate eye protection in the form of a welding mask. You will also need gloves or gauntlets, a chipping hammer, stout footwear. Welding in your shorts, no matter how warm may well result in burns and a fetching instant suntan! It is advised that you attend at least a one day Beginners Welding class or workshop, where you will gain a vast amount of knowledge not only about the practice, but about equipment also.

Small power tools, hand tools and basic gardening tools also need to be purchased, and a wheel barrow (*brouette*) is a necessity, especially if you have poultry or livestock. Plastic ones with wide tyres last longer than thin metal ones. Try and buy one with a solid tyre as punctures and deflations are a nuisance to resolve.

MAINTENANCE & SECURITY

So, you have spent a fortune on all these things. You also need to look after them. Set up your tool shed or barn up properly with racks for storage off the ground. Always clean your tools before you put them away, and sharpen and wipe with an oily rag before winter sets in. Wiping wooden handles with linseed oil keeps the wood good and discourages wood boring insects.

Ensure you have sufficient light in your shed to be able to see at a glance where everything is. A security light outside the shed with a movement sensor (PIR) will give you peace of mind regarding security. Talking of security, remove the screw fittings on your door hinges of your machinery shed and replace with domed bolts on both parts of the hinge, with the nut on the inside of the building. This makes it much harder for any thieves to gain access. Determined thieves will cut through even large padlocks. Use a UV marker pen and or a Smartwater kit to mark your machinery, tools and trailer with your surname and postcode in a not too obvious place on the item. I would recommend that you keep a folder with photographs, serial numbers and information about model numbers or identifiable marks this will help you to make a list for both the police and the insurance (of course you have already done this for ALL the valuables in your house).

Lawnmowers, strimmers, hedgecutters, rotavators and ride-on mowers all need fuel emptied out before winter or prior to the start of the new season. Check and order any new chainsaw chains at this time, and sharpen blades and then oil. If you can service your own machinery, it will save you a fortune and provide a great opportunity to go play in your shed on a winter's day.

A basic service for a lawnmower, strimmer and ride-on will include:

* clean the equipment
* empty old fuel and old oil
* checking & change or at least clean your air filter
* replace oil with fresh and to correct level
* ensure all parts are freely moving and greased
* sharpen and oil all blades
* remove all grass debris and dirt from mowers and ride-on mowers
* for ride-on's – check tyres or punctures & wear, check drive belts for wear, damage and check the general condition of the deck for corrosion.

Ensure your shed has a First Aid box, and if you weld, an eye wash station, and a burns kit. A CO_2 fire extinguisher is also a good idea or a bucket of sand or pail of water.

Champignons sur pain

Olive oil or butter
2 cloves of garlic, peeled & crushed
300g mixed mushrooms, wiped clean
A little sprig of fresh thyme or wild garlic leaves
Some chopped bacon or lardons (optional)
1 lemon
Salt & pepper
2 slices of pain de campagne (country bread) for each person

In a large frying pan heat the olive oil and the butter. Tear or slice the larger mushrooms into bite-sized pieces. Add the crushed garlic and then the mushrooms to the oil and gently shake to coat all the mushrooms. Add the bacon or lardons now and the thyme or wild garlic leaves & salt & pepper. Add more olive oil or butter if the mushrooms look dry. When coloured, simply heap on top of the bread and add a squirt of lemon juice. Serve immediately.

Chapter 7
Acquiring Skills & Training

This chapter is a really short one, because in the rest of the book I have highlighted where and how to gain experience and what legal certificates are required and how to attain them. Quite simply, for some jobs you will need training, some jobs require certification and some need hands on experience. If you have been a registered electrician or plumber in the UK, do not assume that you can bring this certification to France and legally be able to either work in France without taking a further French qualification; or that you can legally undertake work on your own property. Physically yes, you probably can, but any works done may not be recognised and you could invalidate your house insurance.

The easiest thing is to find out before you move what the requirements are and then decide if and when you can acquire that training or certification before you move. There are a wealth of social media sites that will help you to understand what is required and give you links to official French sites. I use Google Translate for most documents and also ask experts in the field. Information is power and knowing where to search for information is key. I have highlighted a few areas below where basic necessities are singled out.

Some of the simplest ways to learn things can be to utilise social media. Yes, it can be full of nonsense, but the gems are there too. In the text of this book I have highlighted the benefits of YouTube for short (and longer) videos showing you, for example, how to prepare a rabbit for the table, but the scope is huge. Facebook has many useful pages and "English speaking" local groups that could be a help for you to find services and properties, how to find good estate agents and more specific help such as how to change your UK driving license, register your imported car etc.

You cannot hope to find everything you need to know at the one time. Set yourself something to find out daily with perhaps a weekly theme. This will hopefully prevent you being overwhelmed. Remember, everything you know today was learned over a lifetime. I'm happy if I learn to pronounce a new word or phrase correctly in a day!

I have also included in this chapter some case studies compiled from smallholders and people who run small rural businesses here in France. The variety and experiences of each should indicate that rural enterprises and ways of life are as individual as the people who run them.

LIVESTOCK TRAINING & HANDLING

If you intend to keep livestock of any type, it pays to get some experience with working and handling them. You don't need to undertake a certificated course, but many private UK providers and agricultural colleges offer day or weekend courses for the aspiring smallholder. It's not just a case of reading about their needs and requirements, about breeding etc, you will need to learn how to handle these animals and also check them for injury. Handling and moving large animals such as sheep, goats, pigs and cattle takes a lot of physical strength, and it is better to find out you are not up to the task before you purchase them.

Prospective poultry keepers would benefit from making time to learn how to dispatch and cull birds, as in France you may wish to remove old or sick birds or prepare them for the table. If you wish to keep sheep or goats, learn how to trim feet and catch and restrain animals for vaccination, worming and if necessary shearing or to assist animals with difficult births.

If you want to milk goats or cattle, then do a day course or go and volunteer on a small scale dairy enterprise! Learn as much in advance and ask the provider if you may email them in future if you come across a problem. Networking and building up a range of friendly contacts will build confidence and reduce any anxiety you have as a beginner.

Many local beekeeping clubs and associations in the UK offer short experience days and longer training sessions and even team you up with a mentor to see you through a year beekeeping. I did this and can recommend it to any aspiring beekeeper. There are now businesses in France also offering beekeeping experience days and will tell you all about what you need to do to keep bees in France. Many people who think they want to do beekeeping will admit after a few times helping at the hives that it's unfortunately not for them. There is no shame in this. Better to try it and retire gracefully from the field than to buy all the equipment and bees and then discover that you are out of your depth.

When in France, try and buy from a local farmer. This will help build a bond and hopefully your new contact will be able to help you with any issues you have initially. Perhaps this may be the start of a great friendship. My farmer friend Pierre who lives in my village has turned out to be a great friend in many more things than just animals. I'm hoping he will offer to take me mushrooming this autumn to help me identify which I can eat and which to avoid and thus prevent me poisoning myself and John.

Looking at what existing skills you can bring to the party might seem like a futile exercise, but we all have a past and skills picked up along the way. For example if you've previously worked as a mechanic or in engineering you will have some existing knowledge of how machines and engines work. You might be able to service your own car. With some more training or research you might be able to service a small trailer. Look at your existing skills as a stepping stone to learning new ones!

A good cook can learn how to preserve fruit and vegetables, and with some research could easily lean to make liqueurs, juice & cider. You might be a computer wizard who can make and update websites, integrate social media and explain to us not so technical people how to do all these things.

If you intend to set up any sort of business in France, including *gites* or producing food or drink, you really need to be business-minded. A SWOT analysis (Strengths, Weaknesses, Opportunities and Threats) will help you identify your skills, strengths, weaknesses and is a precursor to perhaps writing a simple business plan. On an A4 page, divide the page into 4 sections with a couple of dividing lies. In each 'box' write the words Strengths, Weaknesses, Opportunities and Threats. Start to list under strengths all the things you currently have and are good at.

For example, under Strengths, my list included: good communication skills, animal husbandry knowledge, some website admin knowledge, horticultural knowledge, teaching qualification, First Aid qualification. It also contains physical attributes, both of YOU as an individual and of your business. So I've also listed good health, physically fit, good transport links, outbuildings, supportive partner. You can add to this list as you go on.

Under Weaknesses, I've put small plot of land, very small

Capital investment, children, managing mother with dementia and time management. These are the things that are going to hold me back or that I need to address in order to progress. Some things you will have no control over – just accept them. Then complete the 'boxes' for Opportunities and Threats.

Opportunities may only present as your knowledge of your chosen area grows. For instance, you might list that you will be the only provider of organic local apple juice in the area. You might further add that you already have an established suitable orchard. If you intend to run a *gite* your USP (unique selling point) might be a heated pool (the only *gite* property with a heated pool in the area) which allows you to extend your season by 4 months and also elevates your gite to the luxury market.

Under Threats you need to list your competitors and things that could affect your performance (i.e. prospective changes in the law or fashion fads). Also list if inadequate funds will prevent you from expanding or renovating your business or prevent you marketing extensively. This year, package holiday providers have had a difficult time and consumer confidence is now low. This will affect the hospitality industry in France and customers may be encouraged to book direct rather than use an online or traditional high street travel agency and to take adequate travel insurance as a result, so this could actually be an opportunity and not a threat. It all depends on how you look at things.

BOOK-KEEPING

I was always rubbish at maths in school, but I'm pretty good with money. Once I found a simple way to set out my book-keeping, I could easily manage to do business book-keeping for my previous businesses. I've also been on one day courses to fine tune how to set out accounts and learned little tips to help me cross check at a glance, and, let's face it; one day is more than enough unless you really are into figures. A long time ago in the UK, when I was setting up my first business, the government was offering start-up training for new businesses. Adult Education is another place to learn basic business knowledge.

The type of business you intend to have will determine how simple or complex you need your book-keeping to be. Spend some time with a business book-keeper or accountant to learn how to set up your records and how the French authorities like them presented. Or employ a book-keeper or accountant for a few years until you get into the swing of it. Keep copies of all records, and photocopy all returns and keep these for a minimum of five years; the legal requirement in France.

MARKETING & COMPUTER SKILLS

Internet skills in this age are a must and constantly changing and evolving and sadly, some of us are falling behind in this. I had existing basic computing skills but needed to increase my knowledge of social media and how to use it, and then after learning the basics here my confidence grew and I realised I needed to up my game by building my own website. Again, I went on a week long, fully funded course in Cornwall. My head was aching by 4pm daily, but by the end of the five day course I managed to build my own website using Wordpress and understood the benefits of it and how to manage and update it. It's not a complicated website but it works and I can reach so many more potential customers. Ensure your website displays properly on a smartphone, some website templates don't.

Of course, once you learn one thing you realise you need

to learn another. After learning to make my own website, I needed to learn more about social media. This year I'm trying to learn more about search engine optimisation. It's not about learning something for the sake of it, but finding tools to help me manage my business and which will hopefully reduce work further down the line for me.

Try and learn something new everyday and don't expect to be a genius in a week, but one day things just click and it makes sense. To learn how to use Instagram, I had an informal morning course in a coffee shop in Truro with a lady who took the time to sit with me and show me how to set it up, how to use it and why it would expand my business. It cost me £50, and a coffee and a cake. My son was shocked when I told him, but he refused when I had previously asked him to teach me.

Of course, a clever person would also be looking at other people's skills. Why would I learn how to do something complicated when I can pay someone a modest sum to undertake for me? The first time I engaged an accountant in the UK, my first question was "If I use you, how much money can you save me?" He took my tax return, my bookkeeping yearly summary and within a half hour told me he could save me £200 more than the fee he would charge me!

Case Studies

I have included here some useful and varied case studies, put together with the help of some generous people who have allowed me to share their own personal experiences of moving to France. Some have given quotes and some tips as to what they think a new migrant would need to know. Some run businesses and some have just moved here to start a new life for themselves. I'm sure you will find all of them interesting.

CASE STUDY   Sandra Arscott Kempster
Rancon (87)
FB page   Plants and shrubs for sale Limousin

Sandie & Kim made the decision to leave the UK for France very easily, having no real ties to cut. They sold up in the UK in their late forties/fifties and bought an empty barn to convert. They didn't have much money and knew they had to find work within a few months of settling permanently. Kim started doing some long distance lorry driving and Sandie managed to secure some garden maintenance work, which lasted almost nine years. More work started to come in and between times the couple renovated their barn, which is now a small cosy house.

As Kim was formerly an electrician builder he saved a fortune on artisan fees. Although Kim is now semi-retired and awaiting a shoulder operation, he continues to work as a builder. The garden of the house is the 'shop window' for Sandie's plant nursery business. She has now given up garden maintenance work but increased both the propagation of plants and buying in from suppliers for a steady stream of customers. Facebook is her main method of advertising and keeping in touch with customers old and new.

Polytunnels help with the production and care of the plants, but are at the mercy of strong winds and also in high summer need constant watering and ventilation. There has been a reduction in the plant sales due to nationwide watering restrictions in summer in France, but Sandie and Kim are weathering this as customers are regular and loyal. Coffee mornings monthly allow like-minded people to meet at the nursery and Sandie offers sound planting and growing advice.

Quote: "Getting to know the 'right' people is important as there are good and bad everywhere. We came here to integrate with the French and have learnt French; we know at our age we will never be fluent but we have many French friends who we socialise with. You really need to try otherwise it can be a lonely place, couples in particular need to be aware of this in case they lose their partners, and a car is a must also. Lunch shop closure between 12pm and 14.30pm can be very frustrating; as is the paperwork. I often threaten to employ a secretary to do all our paperwork as there is no end to it, when time could be spent doing so many other things and getting on with life you go to your post box and there will be something else to deal with. So if you think you can deal with the above two problems, give up your Cheddar cheese, English bacon and eat good French food at French restaurants with the odd smelly French toilet you will be halfway there to enjoying your life in France. Believe me it is truly worth it, we would never return to the UK, we have been welcomed here and this is our home."

CASE STUDY   Melody & Martin Berry
Ferme du Berry
Champsac, Haute-Vienne, Limousin (87)
www.berryemporium.com
Email: enquiries@berryemporium.com

Melody and husband Martin, lived in a rural setting in the UK, both having fairly mainstream jobs. For Melody, the love affair with France began long ago as the result of a school trip. Illness and pressures of work were the catylists to sell up in the UK and move to France in their mid years, thus freeing themselves of a crippling mortgage. Like most things in life, there just seems to be a 'right' time and 2019 was their time. The move has proved to be just what they need – motivation to start a Good Life for themselves in a more supportive culture. Luckily Melody was already pretty confident speaking the French language.

When they arrived at Ferme du Berry in 2019 they applied form a Cheptel number, and within months they upgraded to Siret d'Agriculture number. This allowed them to legally sell their farm produce. The property is a traditional farm within a hamlet on 1.86 hectares (4.5 acres) of land. The livestock consists of alpacas, Angora goats (both kept for fibre), and some Saanen dairy goats for milking, poultry and ducks. There is an acre for fruit and vegetable production. Some goats and chickens are destined for meat, so the couple aim to be pretty self-sufficient. The fibre is spun for yarn, which is Melody's department as is birthing time. Martin does the daily animal care. As smallholders any income earned is viewed as a bonus for paying the utility bills etc. They anticipate that 50-70% of their income will come via meat, dairy and some fruit/veg. As the flock of angora grows, they will join a co-operative to take our surplus mohair for additional income too.

A UK pension is supplemented by the growing business. Melody is registered for both the agricultural side of the business and also with the Chambre des Metiers for the crafting side. She has found the various government departments, and the local community to be supportive. She loves the strong community feel and importance of family here. Their health has improved considerably since moving and they are both more active physically.

TIP: Patience is a virtue here. Getting used to the 2hr lunch closing, and no Sunday openings is annoying. Then you begin to understand and appreciate that you are better for the break. Register (free) with your nearest Alliance Pastorale and you can purchase your farming needs at a reduced rate on co-operative prices. Keep your eyes open or ask at Chambre d'agriculture about possible grants. If you don't ask, they assume you already got it covered. If you don't speak French, just do your best. If you need to deal with administration, use google translate to prepare you. Above all, please avoid the naysayers. If it's really as hard as they like to portray, then why are they still here, and why are there so many others who have come over here and made a successful lifestyle?"

CASE STUDY   Eco-Gites of Lenault
Rosie & Simon Hill, Calvados
Normandy (14)
Eco Gites of Lenault http://ecogites.eu
Twitter @EcoGitesLenault
Facebook - EcoGitesLenault

Rosie and Simon were in their forties when they moved to France in 2007 with their two young children from a suburban setting near London. They chose France for 3 main reasons – firstly, to have a smallholding and grow and produce their own food; secondly, to run an eco-friendly gite business and thirdly to run a profitable small business. They wanted the children to have a rural upbringing. As the children were very young they settled well into school, becoming bilingual a few months later. They choose Normandy to be close to the channel ports for British holiday makers wishing to travel ecologically, coupled with the fact that Rosie doesn't really do 'heat' and the Normandy climate suited them better.

The family bought a property and renovated a derelict barn into an eco-friendly gite, which took 3 years by which time the smallholding was up and running. In 2013 Simon set up his landscaping business. The property has 0.6 hectares (1.5 acres), but they also have the use of another field for grazing animals. Poultry, pigs and sheep are the main animals kept, and the family are now self-sufficient in meat, eggs and soft fruit, and over 75% self-sufficient in vegetables.

TIP: Move when children are younger as it is easier for them to pick up French and fit into the education system which is a little more formal than in the UK. That said, it is not impossible to move with older children but that will depend on the individual child and some older children take a school year again (redouble). Move in the summer so they can start school at the beginning of the academic year in September. Remember, too, that if you choose to live in a very rural location it will take longer to get your children to school and all French secondary schools start at 8am and can finish as late as 6pm - add on bus journeys and your children will have to get used to long school days. You might also find yourself driving a lot to pick them up if they finish early or go to sports and clubs.

Definitely do your homework and consider renting for a few months in the area you want to move to. Also visit in winter as France is not all sunshine and blue skies and many people do find it very lonely in what can be long winters. Rural France has much to offer if you are looking for a simpler way of living and are willing to fit into your local community. But be prepared to make mistakes Find out about earning a living, healthcare, schools, taxes, cost of living and familiarise yourself with French bureaucracy. Try and learn some basic French and take French lessons when you are here. And a warning - be prepared to take on difficult-to-navigate French websites and a lot of paperwork!

CASE STUDY Sue Flay
Le Vigeant, Vienne (86)
Sue's Flowers
Facebook: fleursdesue

Sue and her partner moved to France in 2002, firstly living in the Charente. Sue is originally from Plymouth. The initial idea came from a couple of friends and they all moved together to a farm with the intention to have a petting farm and mobile homes for holidays. Although the maire was supportive, local opposition came from British residents. Sadly, due to the pressures from all this the friendship ended.

At this time, there were no handholding services and although Sue had O level French and undertook 18 months of French classes, setting up a business was very difficult. The couple bought a renovation project, then another, and did well until the crash of 2009 when they managed to sell the last renovation.

They then bought a house, but had no land. They bought land with greenhouses on site from a neighbour who suggested starting a plant nursery as they were going to retire. They taught the couple growing and propagating techniques, introduced wholesalers and then Sue and her partner approached the Chambre d'Agriculture and the Chambre de Commerce. They decided to go with the latter as it was a simpler set-up and they are allowed to grow 30% of their plants this way. They are registered as Auto entrepreneurs and as such cannot claim business expenses (i.e. costs of compost, pots etc.).

From a small start in a local market, they now do 4-6 markets weekly and are open at the nursery 1-2 days weekly. They make enough to pay the bills. They now have 2 hectares and 2 polytunnels.

Quote: "I never realised what a massive commitment this was going to be. It is really hard work and very long hours if you want it to support you financially. The weather is mostly against you – too hot and dry most of the summer. I would not recommend this as a job unless you are prepared to put in long hours for a small return. We do enjoy it in a weird way though."

CASE STUDY Moulin de Gô
Moulin de Gô
St Pierre sur Erve, Mayenne (53)
www.moulindego.com

The brains behind the renovations at Moulin de Gô is a shy individual who prefers to remain anonymous. He is now what we would term in his 'golden years' and has been in love with France for over 40 years. After running successful businesses in the UK he bought a large renovation project near Le Mans because it was handy for the ferry ports and also the perfect setting for a holiday gite business; and has completed many building projects before buying a 300 year old derelict watermill in 1989. (The mill was probably in existence beforehand, but first documented in 1772). Renovation of the mill did not start till 2012. He bought and renovated the house he now lives in, which is just 10 minutes from the mill in 2006; and rented a nearby *gite* for six months whilst the majority of work was undertaken by himself and another artisan.

The mill is still being actively restored and located on the River Erve. The high rainfall and extreme flooding has at times hampered the work, but the two main workers are helped by occasional volunteers from the village. The work is taking considerable time to sympathetically restore the building back to its working grandeur will not happen overnight. There is an additional 3 acres of land attached to the mill which is managed by volunteers (Les Amis de Moulin de Gô). The mill has been incorporated into a non-profit making association and the volunteers have really taken the project close to their hearts. Currently, the mill is having some flooring work done, but is already functional and can grind corn and has been open to the public since 2014.

The *mairie* is delighted that a building that was considered dangerous and destined for demolition is now being fully restored and again starting to become part of the community again. When fully restored, the mill will be open to the public on specific open days, when people can see the workings and enjoy fresh baked bread from the mill's bread oven. The villagers also are happy to see the mill being refurbished and hope it brings more visitors.

The restoration won a prestigious heritage award in 2018 and the association can feel rightly proud of restoring this important historical gem. In summer 2020, the association hope to build and populate a sculpture park in the grounds next to the mill.

Quote: "Try hard to integrate with the French. This together with becoming involved in your local community will ease your way into society and acceptance. Oh, and be wary of some of your fellow ex-pats. Some Brits think they are way better than the locals and this does not go unnoticed."

CASE STUDY   Hazel & John Rudge
Cheyroux, Payzac, Dordogne (24)
FB page: cheyrouxdonkeys/

Hazel and John moved from Worcestershire to Dordogne in 2012 with their 16 year old son, dogs, horses and two donkeys. They had never visited France before and were about to embark on a massive adventure, but although the purchase of the French property sailed through, the sale of their English property fell through; meaning Hazel moved to France , leaving John to live for a further 10 months in the UK until their house sold. Their son, originally very keen to move to France ended up hating it and moved back to the UK and into residential college accommodation.

The new home; derelict, but structurally sound is now a modest farmhouse, with a couple of outbuildings and nestled within 17 hectares of pasture and woodlands. The couple now have a vegetable garden, a small orchard and gardens. A stream supplies one of the horses fields and a natural source in the donkey paddocks is filtered to provide both the house and the donkeys with drinking water.

When searching for a suitable property, the couple knew that having enough land for horses and hay making was essential, as was the land being attached to the property. They wanted a traditional, cheap to update building in a rural setting. Funds were low and to make an income they became auto-entrepreneurs and both now offer part time seasonal gite management, changeovers and home and garden maintenance. John manages the woodlands to provide fuel for heating and also enjoys falconry with his Harris Hawk.

Soon more rescued donkeys came to live with them, and in 2018 Hooves at Cheyroux started almost by accident; offering a forever home to abandoned and mistreated donkeys.. They are in the process of registering as an association, but receive no official funding; relying on the goodwill of animal lovers and their own jobs to fund this.

Hazel found learning French the hardest part of the journey, whilst John struggles with the amount of time it takes to get things done. They are happy to welcome visitors by prior arrangement.

Quote: "Plan, plan and research. We didn't but we were very lucky! Be flexible as you may need to earn money in ways you never imagined. If there wasn't an ex-pat community here we would be in serious trouble financially as they are our work clients. We came with no savings and lost lots on the sale of the UK house which was intended to be our safety net, so we had to hit the ground running. Be prepared to go with the flow as French life moves to a different pace than UK life. Isolation is also a bigger issue here so you need to like your own company if you intend to live rurally."

CASE STUDY   Mas Del Pech
Richard & Tina Holland
Concorès, The Lot (46)

Richard and Tina moved to France in 2013. They previously lived in a 2.5 acre smallholding in Devon for over 20 years where they lived a rural lifestyle with goats, sheep, pigs, poultry & gundogs. Their children are adults and have remained in the UK, but visit frequently.

The decision to move to France came as a result of visiting friends in Brittany. With one son still at home, the couple talked about the possible move, but that was all. As they turned 50, and had no children at home, the recession in the UK hit. Richard was made redundant and had a heart attack and complex surgery afterwards. Suddenly, life becomes more precious, and the couple decided to sell up and move to France. Brittany was too like Devon, but a TV programme featured the Dordogne and the couple liked the look of it. Two visits later and the decision to move was made. They 'borrowed' a friend's holiday home and moved just before Christmas with nine dogs. Finally, after much property searching they fell in love with the Lot. A farmer friend advised looking for a hilltop property for cool summer breezes and little winter frosts and they soon found their home.

Mas del Pech is a small traditional stone fermette of 5.5 hectares (13.5 acres), mainly rough grazing with some apple trees and half a hectare (1.2 acres) of woodland. The couple have a small flock of Causses and Causses cross sheep which they keep for meat and breeding. They also have some chickens and ducks. There is an arrangement with the neighbouring farmer who grazes cattle on some land in exchange for cutting & baling hay for the couple.

The locals have been very friendly and helped them buy sheep, get registered with the Chambre d'Agriculture and settle into hamlet life.

The income is derived from about 50% pension and 50% a mix of Richard's work in a vineyard near Cahors and Tina's soon to be developed dog sitting enterprise known in French as *Le Garde de chien en famille*. As well as getting both necessary permissions and sitting a multiple choice exam (in French) on a computer at the test centre; the couple have had to comply with a variety of conditions including having a designated reception area for the business and ensuring that the business area is at least 200m from the next property.

Quote: "Rent if you can for 12 months and experience all four seasons. Match the area you choose with what you aim to do or produce. Check both the local demand and local markets for your intended business, if you're going to do that and stick to what you know; at least at the start. Officially check all advice – Facebook is fraught with bar room lawyers."

CASE STUDY   Christina Powell
Lascaux Permaculture Gardens
Haute-Vienne (87)
www.lascauxpermaculturegardens.com

Christina and Chris Powell are in their early 40's and moved to their 5 acre smallholding in 2016. Originally from Romania Christina was brought up on a family farm and has loads of experience growing vegetables and keeping animals. Chris is British and the couple formerly lived in the Midlands where they had an allotment.

They chose France as their home to benefit from the space and peace and quiet and also to be able to live a more sustainable life and run a self-sustaining business.

Around half of their land is woodland and just under one acre under cultivation as a market garden. The couple have embraced the idea of spreading risk and produce seasonal veg boxes, herbs and perennial plants, eggs from their large flock of hens, and dairy produce from goats and Jersey cow. Ducks, geese and male poultry produce meat for the couple's own use.

Of course the couple are registered with the Chambre d'Agriculture and registered as a Micro BA and found the paperwork side time consuming but not overly difficult. They did some research before making an appointment to discuss their plans before registering. Chris works in the UK whilst the business is establishing itself and Christina runs the market garden almost single-handedly.

The house is an 18th century stone renovation project and this together with the market garden and caring for the animals takes up all the couple's time. Developing dairy products is the next stage and the couple also hope to have pigs soon.

TIP: Set a goal outline, never a goal; because in farming you have to be able to deviate, make adjustments and change course. Always start small so you can thrive in your endeavours.

Quote: "For me, being a smallholder is a unique experience and chance to reconnect with your inner nature's calling; you give your time to animals, vegetables, wildlife, fruit trees and bushes and you get rewarded with bountiful food and unconditional affection from animals."

CASE STUDY Daniel Teare
Le Gast
Normandy (38)

Daniel and his partner Jules are in their late 40's and moved from the Bristol area to live in France in 2014. They had initially been looking to find a smallholding to rent in the UK but came across their property via **www.greenshifters. co.uk.** The decision to rent was as a result of being afraid that they might not be able to resell their property easily. They rent 4 acres which they share with an assortment of animals including chickens and pigs; some of which are long term residents and some which are destined for meat. Their land is 60% self-sufficient in vegetables. Daniel and his partner arrived in France as vegetarians but now Daniel does eat home produced meat. They find meat production, especially slaughter difficult, but accept that as part of the experience of keeping animals and living sustainably.

Having now gained a lot of experience in smallholding they have decided to return to the UK to start another chapter in their smallholding journey.

The aim of coming to France was not as a money making exercise, but as a lifestyle choice and they have learned a vast amount here and lived comfortably with little income. They have been accepted well by their community and have learned French whilst living and working on site. They have found the wet winter weather in Normandy trying for their smallholding lifestyle.

Quote: "The language has been difficult but in general it has been all good. The best tip I was given by a neighbour when we first arrived was you can sleep with Frenchman's wife but don't touch their fire wood.

Regarding what skills are needed I don't think you need any just work with nature not against it. If you're not fit you soon will be. If there is anything you don't know how to do - google it. It's good. It's all good."

Sauce tomate (basic tomato sauce)

2 tbsp tomato puree
4 cloves of peeled & crushed garlic
2 onions or shallots, peeled & chopped fine
4 tbsps of olive oil
A bunch of fresh basil leaves, torn.
A sprig each of fresh thyme & origano
800g chopped plum or meaty ripe tomatoes
Salt & pepper

This makes around 500ml of tomato sauce and can be frozen. In a large frying pan fry the crushed garlic with the onions or shallots in the olive oil till translucent. Add the tomatoes and the herbs and stir. Add and stir in the tomato puree. Season with salt and pepper and bring to the boil, stirring all the time. As soon as it starts to boil, remove from the heat. Strain through a coarse sieve and discard the herbs, any garlic remains. Return the sauce to the saucepan, bring to the boil again, then turn down and simmer or 5 minutes to thicken slightly.

Chapter 8
Making It Pay

Registering any business and dealing with the correct taxation is a complex area and you are strongly advised to enlist some expert advice before you register anything. There are many English speaking experts in France who offer handholding services, taxation, bookkeeping, translation and legal advice.

Depending on your own circumstances and choice of business you could consider registering as a micro-entrepreneur (a sole trader), *travailleur indépendent* (self-employed individual). But you MUST take advice as to which regime is best for your circumstances. There are 3 main categories of self-employment in France; as a *profession libérale* (e.g. a doctor, accountant, writer), an *artisan* (craftsman or tradesman) or *commerçant* (trader or shopkeeper).

Setting up in business in France is very different from setting up as a self employed person in the UK, as there are tax implications and social security implications for different categories, and this is where a hand holding service or fluency in the language are important to get the right information before you register. You may have to undertake a course run by the appropriate organisation, and most courses are run in French. The system has been somewhat simplified recently and it is certainly possible to register as a micro-entrepreneur online for a small fee.

If you intend to register as an *agriculteur*, (farmer) please take advice as to registration type as the registration type. The registration for someone keeping sheep for meat is very different from keeping sheep and producing wool, for example. If you intend to keep a small amount of animals for your own use only, and not for commercial gain, this is another option.

HOLIDAY ACCOMMODATION

Whilst this may at first glance appear to be the easiest and surest way to get your property to cover the cost of you living in France, please do spend some time and look at things in detail before committing.

Firstly, in the more touristy areas of France the market has become saturated with *gites*, bed and breakfast and campsites in the last 10 years or so. The rise in popularity of AirBNB has meant that some accommodation providers simply didn't register with the French authorities and this lead to prices being forced down by providers cashing in without paying any tax. Naturally, this situation was never going to last and the law now states that ANY accommodation must be registered with the *mairie* and with the tax authorities, whether you are resident in France, UK or elsewhere.

Look seriously at the season you can expect to fill. For example, if you are looking to buy a place in a popular area for skiing, then you may also be able to offer summer accommodation to walkers, cyclists and photographers, but you are not looking at all year round occupancy. At most you will be able to easily fill 4 weeks over the Christmas holidays and possibly the occasional week or weekend in winter, and perhaps some weeks in summer.

If you intend to purchase in the South of France on the

Mediterranean coast then again, you can almost guarantee to fill July and August, but may also get an older population booking on shoulder seasons too; if your property is accessible for older people and is near to an airport.

If you are looking to market to businessmen and commercial travellers, you will need to provide cheap, cheerful accommodation with excellent internet provision near to main road networks; and be prepared for one or two night stays, which are a lot of work and not much profit.

Providing B&B (*Chambre d'hôte*) with breakfast is a lot of work, unless you either provide an open kitchen with continental breakfast and a coffee machine, or offer a high end breakfast with more choice. If you also intend to offer an evening meal you will have to consider guests with food allergies, requests for a change to the menu and a much longer working day. If you offer hot food you must also check the requirements to attend courses before registering your business. Running a *gite* is far less work as your guests are self-sufficient for their food, and you can offer a minimum stay of 3 nights off season and minimum 7 nights in high season.

Ensuring you get guests to pay a non returnable booking deposit when securing their stay will prevent a lot of cancellations, and taking a returnable damage deposit alongside ensuring guests agree to a set of terms and conditions will go some way to protecting your bookings and property. There are many Facebook groups offering help to new hospitality owners including Gossip for Gite Owners. **https://www.facebook.com/groups/GiteOwners/**

Developing an area of land for a campsite or caravan park may present itself as an option, or perhaps you are considering buying an existing one. Firstly, look at the existing provision in the area. If there are already more than a few good ones, then ask yourself what YOU could offer to attract those holidaymakers. French sites tend to be excellent, with multiple toilet and shower facilities, at least one children's pool and an adult pool, and often have on site snack bars, restaurants and a bar. With the investment in facilities required, such sites tend to be huge with many chalets, motorhome hook-ups and tent pitches. Are you really prepared to compete with this, especially as the main letting period is July and August only? And how many staff will you need?

How are your marketing skills? Can you put a website together and maintain it, social media and the various online travel agents popular to market your accommodation business. Prepare to work on your marketing and answering queries for at least 2 hours daily every day, almost all year round. For my own *gite* business my enquiries start coming in the first week in January and don't stop until around November.

Think about your USP. Your Unique Selling Point may be a heated, covered pool (allowing you to take guests all year); it may be your position overlooking the square in a medieval town or city. It is what makes you more attractive than your nearest competitor. Offering secure cycle storage and good information on nearby cycle routes may make you popular with cycling tourists as will proximity to the route of the Tour de France. Cyclists tend to holiday outside the main summer season, and this could be a good market for *gites* and or *Chambres d'hôtes*. See: **www.francevelotourisme. com** for cyclist tourist information.

KENNELS, CATTERIES & PET SITTING

Most French residents tend to take their pets with them if they holiday within France, and many *Chambres d'hôtes* and gites accept pets. However more are looking at pet day-care and accommodation and there is an increasing demand for the service, especially in big cities. Many British ex-pats find themselves having to use kennels and catteries (or house-sitters) if they have to return to the UK as it's simply too expensive or inconvenient to take pets with them.

There is more kennel provision than cattery provision, but both could offer a lucrative business opportunity, *if* you have sufficient outbuildings and are in a semi-rural or rural situation where noise won't be an issue, and can fulfil the necessary requirements. Your property must be at least 100 metres from any neighbouring property if you plan on housing dogs. Good fencing will be necessary, and of course showing that you have made adequate provision for the well being and health of any animals, including dog poo bins and how you intend to dispose of dog waste. Your first port of call should be your *mairie* to enquire about provision and whether you would require a change of planning use to commercial (a *déclaration préalable*).

You have to be approved by your prefecture or by the departmental veterinary service (DDVS) with whom you require to be registered. Before registering you must have obtained the *Certificat capacité des animaux de compagnie d'espèces domestiques* (ACACED) for dogs or cats.

This may depend on what sort of arrangement you wish to have for animals boarding – whether boarding in your home (*Garde de chien en famille*, or *pension familiale pour chien*), or as kennels or a cattery. The course has a long waiting list, costs around 280 euros and consists of a 2 day course ending with a 30 minute exam in French. More details here: **www. formation-certificat-capacite-domestique.fr/**

Dog walking and house sitting is an increasingly popular opportunity in cities and for business owners who work long hours. Trusted house sitters can provide good references and must have experience with dogs, cats and possibly horses or chickens.

A step up from dog sitting and kennels is the provision of livery stables or a riding school. Again, contact your local *Chambre d'Agriculture*, and ask a local or specialist insurance company for advice, as you will almost certainly require public liability insurance.

FISHING, LAKES & RIVERS

France has some beautiful countryside, often with streams, rivers and lakes. Many of these are teeming with freshwater fish, and some lakes have been specially provided as fishing lakes. You require a fishing licence to fish anywhere, and the licence may be granted by a local fishing club private landowner or by the local community. A *carte de peche* must be purchased before fishing and you have to decide in advance what species you are trying to catch.

Day or weekly cards are available which allow short periods of fishing, usually between June and September. For more information have a look here: **www.cartedepeche.fr/215-liste-des-cartes.htm**

If you have some experience of coarse fishing you may wish to exploit your own lake if you have one and could offer camping or caravanning and fishing would be your USP. If you intend to offer commercial fishing the lake should be officially registered and you will also require to register with the local Chamber of Commerce. Lakes with closed status, i.e. have no stream entering or leaving the lakes do not require annual inspections or water tests. If you intend to buy an existing commercial fishing lake, then do your homework and check that all the registration etc is in place before proceeding.

SELLING EGGS, MEAT AND OTHER FOODSTUFFS

If you run or intend to run ANY commercial agricultural enterprise (i.e. you are being paid) in France then you must at the very least join the MSA (*Mutualite Sociale Agricole*) and pay monthly cotisation until the business ends. Think of this department as an amalgamation of Social Security and the Farmers Union (NFU). It is the equivalent of CPAM for Micro Entrepreneurs. This can be very costly and time consuming to arrange and complete. Unless you are very keen or very sure that you are going to make a lot of money doing this (whatever your money making project) then the general advice from people who have tried before is to think again and just produce for yourself or for friends. You also cannot sell products from your farm outside an 80km radius of the farm.

For example, to sell eggs at a market – you need to register with the DDCSPP and the Chambre d'Agriculture. Depending on the scale of your business you may register as a Micro BA (Benefices Agricole) regime, but please research this. You will require an egg stamp with your individual number, an ink pad and supply of food safe ink. All eggs must be candled to ensure they are uncracked and have no embryos growing inside and are free from contamination. On the box you must state the date of lay, the consume by date (3 weeks from Date of Lay). It costs approximately €700 per annum for certification and you will be inspected annually.

If you wish to go a step further and perhaps sell oven ready poultry, then bear in mind the cost to produce these birds. Even fast maturing meat birds will eat a lot of cereal, and you need to balance the cost of rearing, feeding, butchery and selling against the price you charge. The French particularly prize free range produced poultry, but even so the profit is minimal unless you are large scale. Aiming at larger animals such as sheep, pigs or cattle for the production of meat offers a higher profit, and a ready market. Selling to friends whilst simple is technically not legal unless you have registered with the appropriate authorities.

Selling home grown fruit, vegetable, firewood, apiary products or mushrooms may also come under the Micro BA scheme, but you will need to research this. As an aside, and to show how complex the registration system and regimes are, gardening comes under the Micro Entrepreneur umbrella, but must be registered as a secondary activity with earning restrictions.

CRAFTS

Weekly markets abound all over France and your local village or town will undoubtedly have one. If you wish to set up a stall and sell any crafts you make, you need to go speak to your *mairie* first. If agreeable, they will tell you the size and cost of a pitch and any regulations you need to abide by. There is frequently a waiting list, and you will need to be prepared to attend every week and arrive and set up early in all weathers.

Some crafts that do well at markets include basketry, wood turning, stained glass pieces, jewellery and blacksmithing. You still need to be registered with the Chambre des Mètiers et Artisans and at least be a microentrepreneur.

TEACHING

Language teaching, in particular English language is a significant earner for many native English speakers living in France. If you have a good command of French language sills you may find yourself in demand by new expatriates wishing to learn French privately, French students wishing to improve their English and this can be offered full time, part time or during the summer holiday period. A qualification is not always required, although you should be prepared to show how highly educated you are by producing any degrees or certificates you have.

Having a TESOL, CELTA or similar qualification in teaching English will certainly open more doors, but the course is fairly expensive. If you intend to teach in a school or recognised further or higher education establishment in France you must be a French citizen and possess French teaching qualifications.

Teaching other subjects, such as animal husbandry, photography, basketry, cooking or wood working skills will be relatively informal as long as you are teaching adults and not offering any qualifications as a result of the teaching. I have learnt through experience that ensuring all customers book and pay in advance for any courses safeguards any arrangements you have to put in place before the commencement of the course.

Finally, If you are offering any course that involves animals, machinery or the great outdoors ensure that you complete a risk assessment for any and all activities to safeguard yourself and show due diligence. Check what the legal situation is regarding insurance for courses where people are paying for instruction.

PHOTOGRAPHY

The market for beautiful photographs is a wide ranging one. Many magazines and books demand high quality photographs but are not always so keen to pay good money for them. If you have a known reputation and a good contract you will be more popular, but some photographs are sold to magazines 'on spec'. When selling to magazines or publishers, ensure you are clear about whether they are buying a single use of the image or the copyright to the photograph. Ensure you have a well laid out contract to protect your rights.

Travel magazines, property magazines, and special interest publications on cycling, canoeing, skiing are all looking to fill issues with top quality outstanding images. An interesting market could also include book jacket covers and internal photographic illustrations.

Other photographic options are wedding photography, child portraiture, graduation photography, food and restaurant photography and the growing market of website photography and video. The use of drone videos has revolutionised the property and holiday accommodation market and having a short video online now makes an impressive showcase. If you can produce top quality video material you can find a ready market here in France. A picture can paint a thousand words, but a video can paint a whole lot more.

Having an online presence to showcase your work needs to be professionally managed with an online shop and image protection for your images. Large markets and craft fairs are also good places to sell unframed, but mounted images from postcards to A4 and up to A3 size. Naturally, if you are in a market or craft fair in a particular medieval town, people will be expecting to see traditional cityscapes and fields with sunflowers and local houses and not seascapes which would sell better at similar venues in Brittany or the Côte d'Azur. Explore your market, do a little market research and investigate your competitors. What is selling, what sort of price and is your own product better?

WRITING

If you think you have a flair for writing and would like to try to make an income from this, then my advice would be to hone your skills by writing short articles for magazines before attempting a full length work. When I started writing it developed from teaching day courses. My first article was about keeping chickens and was published in a British smallholding magazine. I remember it took me around two weeks working on it to get the right balance of humour, interest and fact, and then I sent it with two high definition, large format photographs to the editor with a covering letter.

The article was published and I continued to write occasional articles for a couple of magazines on subjects that were slightly unusual but of interest to the readership. Payment usually arrived a couple of months later, but I have to say that these days payment is small and editorial budgets are very tight.

Possible avenues to reach out to are travel and lifestyle magazines, special interest magazines such as food, wine, photography or transport (if you have expertise in these areas). The old maxim 'write about what you know' is king here, and you have to offer something new to readers and write in an entertaining and informative style. Be prepared for rejection and be ruthless with your spelling, punctuation and grammar. If you can accompany such articles with your own professional quality photographs you will present a more compelling package, but have a look at the magazine's editorial section for their requirements.

In France there are other opportunities ranging from writing promotional copy in English for French tourism businesses or websites (if your French is good enough) or for UK magazines looking for French lifestyle or travel articles. You will need good, clear high resolution photographs to accompany any article, but do not send unsolicited material. Send an introductory email with your proposal and any published material you already have.

Having a book published is every writer's dream, and although you may work for many years on such a project the reality is that after printing costs and advertising etc, writers achieve very little profit if you go down the traditional route of using a publisher. Unless you (and your publisher) are convinced you have written the next Booker prize winner or a scorching sexy beach bestseller then be prepared to see your book at number 1620 on the travel book bestseller list and rejoice in the fact that you completed it and got it published at all.

Self publishing has evolved from vanity publishing and is no longer looked at with derision. In fact, if you have the ability to self publish and the funds for the initial set up costs, this can prove the most lucrative method to publish. The risk is all yours and you have to be realistic about your audience size, the genre and the format you publish in (whether digital or paperback). If you have doubts that all this costly up front investment will not realise even the production costs then you will need to take a long look at whether writing is for you.

If you have a really good command and understanding of the French language then translation is a profitable business to be in. I have a British friend who worked for many years in France, then when she returned to the UK to live retrained as a teacher. She taught French class and also undertook French/ English translation for businesses. She has now moved back to France and translates books, property contracts and much business contracts and technical information from French to English and vice versa.

ESTATE AGENCY

When you have been looking for French properties in France via an estate agent or *immobiliére* you may have thought this was an easy job, and perhaps one which you could easily do when you settle in France. Setting up an *immobiliére* business in France requires registration with the Chamber of Commerce (CCI) and the necessity of an annual *carte professionelle* (licence), a very good level of French language skills and is pretty poorly paid. Many larger estate agencies advertise 'vacancies' for English speaking staff, but do your homework and work out if this is a realistic option for you. Frequently, these are commission only positions.

WEDDING SERVICES

In France marriage is a legal and civil ceremony, not a religious process and takes place at the *mairie* or town hall. The official is the local *maire* and the marriage is conducted in a public room. After this legal ceremony some couples choose to have a religious 'wedding' at a church or other venue. Some churches will allow an English speaking priest to carry out a wedding service. Same sex marriage became legal in France in 2013. For non French residents the requirements to allow a legal marriage in France are cumbersome. One of the parties has to be resident in France for at least 30 days prior to the application to get married. In addition to passports, each party must also provide an original birth certificate, certificate of celibacy, an affidavit of law stating that you are free to marry and that your marriage will be recognised in your home country, a divorce decree if you have been married previously, and a prenuptial agreement if you so wish.

Applications are made direct to the *mairie* at least 10 days prior to the wedding date and may require officially translated documents (authenticated with an *Apostille* stamp). If the application is successful you have permission to get married within one year of the date of permission. The cost to get married is around 100 euros, and after you are married you will be presented with a *Livret de Famille* , and a marriage certificate from the *mairie* where you married. Because of the residency restrictions, many couples marry in the UK in a civil ceremony and then choose to come to France for a symbolic or religious ceremony. And this is where the opportunities arise for wedding service and accommodation provision.

Unlike the UK, there are few wedding planner businesses although very recently some entrepreneurs in France are setting up wedding planner businesses. The most well known have to be Dick and Angel Strawbridge who set up a wedding business in the Loire Valley, offering the chance for couples to get married in a fairytale French chateaux, and offered the ceremony, the meal and entertainment for the day. They help the couple organise local accommodation and anything else to make the day go smoothly.

Now, most people moving to France won't buy a chateau, but can offer other wedding services, such as floristry, photography, cake making, catering, mobile wine van, music (harpists, bagpipe players and small string ensembles are popular), car hire, airport transfers etc, but remember the wedding season is a short one and needs booking with deposits well in advance. Larger chambres d'hôtes and gite complexes may cash in with big wedding bookings also.

Canard rôti a la citrouille, pommes de terre et prunes (Roast duck with pumpkin, potatoes and plums)

2 fresh duck legs (or 2 breasts), washed and dried
1 large butternut squash or potimarron, cubed and de-seeded.
Some small and slightly cooked baby potatoes
6 washed & pitted plums
3 cloves of garlic, peeled & crushed
2 tbsp good wine vinegar
Salt & pepper
2 spring onions, roughly chopped
Juice of 2 limes
Preheat oven to 180C/350F. Season both sides of the duck legs with salt & pepper and place in a roasting tray and roast for 30 mins, turning half way through. Drain, sieve and keep the leftover fat (excellent for roasting potatoes).

Take the squash or potimarron and place in a roasting tray with a little olive oil drizzled over. Add the garlic, plums, potatoes, the wine vinegar, salt & pepper and roast for around 20 minutes until the plums and squash have softened, and the potatoes have taken on a little colour. Remove from oven, pour the juices from the pan over the contents, add the spring onions and the lime juice and serve beside the duck leg.

Appendices

Sadly, if you aim to live in France permanently, you will need to prepare for the time when you or your partner dies. None of us know the time or place, but talking about your wishes is a good place to start and having a file of numbers to call in the event will save time and stress later.

Preparing a will is not a pleasant task, but leaves your surviving partner or children with some direction and allows you to consider inheritance tax issues.

If someone dies in hospital, the hospital staff will notify the *Mairie*; if at home, then the undertaker may do this. A doctor will certify the death and issue a death certificate (*déclaration de décès*). As in France, paperwork is king, get at least 10 copies of the original death certificate.

The body can be removed to a funeral parlour or remain at home if this is your wish. Normally funerals take place within six days of death, although this can be extended if family are returning from a distance. The undertaker will advise on what you can choose as to a coffin, flowers, urn, burial, cremation etc. Do not be shy about asking for the cost of all services as funerals are expensive and never under 3000 euros. If cremation and thereafter repatriation is required, you must inform the crematorium in advance.

You must also make an appointment to see your *notaire*, who will, as a government official, want to ensure that the will is followed correctly, and more importantly who is liable to pay inheritance tax if owing.

Useful contacts & links

www.bbc.co.uk/languages/french French language tuition on UK TV

http://britishexpats.com/forum/france-76/ Help for UK expats

www.cheminsdelarose.fr Roses for sale in France

http://completefrance.com Complete France information & property sales Lots of useful information and property sales.

www.connexionfrance.com English language French newspaper including property for sale.

www.downsizer.net This is a UK based site with a vast amount of information on smallholding, rural affairs and sustainability.

www.expatnetwork.com/moving-to-france Lots of information for new ex-pats in France.

www.thegoodlife.com The Good Life France website and newsletter with a slightly smallholdery theme and lots of general information on France as a country. Janine Marsh's book, My Good Life in France is a personal memoire of moving to France.

www.gouvernement.fr/en/coming-to-france French government advice to prospective immigrants to France.

www.greenacres.fr Estate agent listing properties for sale in France by owners and estate agents.

www.immobilier.notaires.fr Information relating to notaires who sell property.

www.internations.org/france-expats/americans Help for

American expats.

www.jaques-briant.fr French mail order plant catalogue

www.thelocal.fr A good site to use to inform you of upcoming changes which may affect you living in France.

www.maisons-paysannes.org A great site for rural life and gardens in France, including their preservation.

Facebook Group - **RIFT CR - Remain in France Together - Citizens Rights** For up to date information regarding UK citizens rights in Europe and especially France in the event of Brexit.

http://rivierareporter.com English language online newspaper for news and information from the South of France.

Goose Fat and Garlic Strang, Jeanne (2013) Cooking recipes from SW France - highly recommended.

Handholding services

Christophe Dutertre Qualified notaire, dealing in property purchase, tax & inheritance issues.

France Tax Law FB page: **Francetaxlaw**

http://www.francetaxlaw.com

Karen Jones Solutions FB page: **Solutions**

Email: Karen.solutions@outlook.com

Katey Warwick

TheHandHoldingService FBpage: **TheHandHoldingService**

English/French Glossary

A

Acidic : acide
Alkaline: alcalin
Algae: les algues
Aphid: le puceron
Apple (malus pumila): le pommier
Ash: le cendre
Aubergine (solanum melongena): aubergine

B

Bay (laurus nobilis): laurier
Bean (phaseolus vulgaris):haricot
Bed: plate-bande
Beech (fagus sylvatica): hetre
Bindweed (convolvulus): liseron
Blackberry or Bramble (rubus fructicosus): ronce
Blackcurrant (ribes nigrum): cassisier
Blackthorn or Sloe (prunus spinoza): prunellier
Blanch: forcer
Blossom: fleurir
Bog: marais
Bonemeal: poudre d'os
Bordeaux Mixture: bouillie bordelaise
Broccoli (brassica oleracea):broccoli
Bud: bourgeon

C

Carrot (daucus carota): carotte
Cauliflower (brassica oleracea): Chou-fleur
Celery (apium graveolens): celery
Chainsaw: tronçonneuse
Chalky: calcaire

Cherry (prunus cerasus): cerisier
Cherry-laurel (prunus laurocerasus): laurier-cerise
Chestnut (aesculus hippocastanum): marronier
Chives (allium schoenoprasum): ciboulette
Clay: argile, sol argileux
Clematis (clematis): clématite
Clover (trifolium): trefle
Cold Frame: chassis
Compost: compost (er)
Coppice: receper
Coriander (coriandrum sativum): coriandre
Couch Grass (elymus repens): chiendent
Courgette (cucurbita pepo): courgette
Crown (of a tree): couronne d'un arbre
Currant (ribes rubrum): groseiller
Cut (v): couper
Cuttings (v) (to take): bouturer

D

Damp: humide
Dandelion (taraxacum officinale): pissenlit, dent-de-lion.
Daisy (bellis perennis): pâquerette
Deciduous: caduc
Dig (v): becher, creuser
Disease: maladie
Dock or Sorrel (rumex acetosa): osielle, patience
Dog Rose or wild rose (rosa canina): églantier, rosier de chiens

E

Elder (sambucus): sureau
Elm (ulmus minor): orme
Estate Agent: immobilier
Evergreen: plante/arbre a feuilles persistantes

F

Fallow: friche
Fan shape: palmette
Feed (v): nourrir
Fence: clôture
Fertilizer: engrais, engrais minéraux
Fig (ficus carica): figuier
Fir (abies): sapin
Fly: mouche
Fork: fourche
Fungus: mildiou, champignon

G

Garlic (allium sativum): ail
Gerkhin (cucumis sativus): le cornichon
Gooseberry (ribes grossularia): groseiller à maquereaux
Gorse (ulex europaeus): ajonc
Graft (v): greffer, greffe
Grass: herbe
Gravel: gravier, gravillon
Greenhouse: serre
Grow: pousser
Guelder Rose (viburnum opulus): viorne obier

H

Hardy: rustique
Hawthorn (crataegus monogyma): aubépine, épine blanche
Hazel (corylus avellana): noisetier
Heather (erica): bruyère
Hedge: haie
Hedgetrimmer: taille-haies
Hoe (v): biner, sarcler
Holly (ilex): houx
Honeysuckle (lonicera): chèvrefeuille
Hornbeam (carpinus betulus): charme

Horsetail (equisetum arvense): prêle
Hydrangea (hydrangea): hortensia

I

Iris (iris): iris
Ironmonger: quincaillier
Ironmongery: quincaillerie
Ivy (hedera helix): lierre

K

Kale (brassica oleracea) chou frisé

L

Laurustinus (viburnum tinus): laurier tin
Lavender (lavendula): lavande
Lawn: pelouse, gazon
Lawnmower: tondeuse
Leaf: feuille
Leek (allium porum): poireau
Lime (mineral): chaux
Lilac (syringa vulgaris): lilas
Lily (lilium): lys
Lily of the Valley (convallaria): muguet
Lime tree (tilia): tilleul

M

Mallow (malva sylvestris): mauve
Manure: fumier
Maple (acer): erable
Marigold (tagetes): souci
Marjoram (origanum majorana): marjolaine
Marrow (cucurbita pepo): courge
Medlar (mespilus germanica): neflier
Melon (cucumis melo): melon
Mimosa (acacia): mimosa

Mint (menthe): menthe
Mow: tondre, tondre la pelouse, tondeuse a gazon
Mulche (v): pailler
Mustard (sinapis): moutarde

N

Nasturtium (tropaelum majus): capucine
Nectarine (prunus persica laevis): brugnon
Nettle (urtica): ortie
Neutral: neuter, basique
Nigella (nigella): nigelle
Nitrate: nitrate
Nitrogen: azote
Nursery: pépinière

O

Oak (quercus): chene
Oleander (nerium oleander): laurier rose
Olive (olea): olivier
Onion (allium cepa): oignon
Orchard: verger
Oregano (origanum vulgare): origan
Organic: biologique

P

Pansy (viola): pensée
Parsley (carum petroselinum): persil
Paving: dallage
Peach (prunus persica): pêcher
Pear (pyrus communis): poirier
Perennial: vivace
Pergola: pergola
Pine (pinus): pin
Plot: terrain
Plum (prunus domestica): prunier

Pond: mare, étang
Poplar (populus): peuplier
Poppy (papaver): pivot, coquelicot
Potato (solanum tuberosum): pomme de terre
Prune (v): tailler, émonder
Pull out: arracher
Pumpkin (cucurbita mamima): citrouille, potiron

Q

Quince (cydonia vulgaris): cognassier

R

Radish (raphanus sativus): radis
Rake (v): ratisser
Rambler: remontant
Raspberry (rubus idaeus): framboisier
Rhubarb (rheum rhabarbarum): rhubarbe
Ride-on mower: tondeuse auto-portée
Rocket (eruca sativa): roquette
Root: racine
Rootstock: porte-greffe
Rose (rosa): rose
Rosemary (rosmarinus officinalis): romarin

S

Sage (salvia): sauge
Scion: scion (in France this is a year old grafted tree)
Septic tank: fosse septique
Shade: l'ombre
Shallot (allium cepa aggregatum): échalote
Snowdrop (galanthus): perce-neige
Smallholding: fermette
Spade: la bêche or la pelle
Spindle (euonymus europaeus): fusain d'Europe
Spruce (picea): épicea

Straw: pailler
Strawberry (fragaria): fraisier
Strimmer: débroussailleuse
Stump: souche
Sunflower (helianthus annuus): tournesol
Swede (brassica Napo-brassica): rutabaga, chou navet, chou de suede.
Sweet chestnut (castanea sativa): châtaigner, marron
Sweet pepper (capsicum annuum): poivron

T

Thin out: éclaircir
Thistle (cirsium): chardon
Thuja (thuja): thuya
Thyme (thymus vulgaris): thym
Tomato (lycopersicum esculentum): tomate
Trailer: remorque
Train (v): palisser
Trunk: tronc
Turnip (brassica rapa): navet
Twig: brindille

V

Vegetable: légume
Verbena (verbena): vervein
Vine (vitus vinifera): vigne
Virginia Creeper (ampelopsis veitchii): vigne vierge

W

Wall: mur
Walnut (juglans regia): noyer
Water (V): arroser
Waterlily (nymphea): nénuphar
Weed: la mauvaise herbe
Weed (V): désherber

Weedkiller: désherbant
Wheelbarrow: brouette
Willow (salix): saule
Wisteria (wisteria): glycine
Woodburner: poêle à bois

Z

Zinnia (zinnia): zinnia

Index

A

B

Printed in Great Britain
by Amazon

47118033R00118